NEED WATER

THE PREPPER'S SURVIVAL GUIDE

A practical guide to find, harvest, purify and store your most essential survival resource.

Home, Off-Grid and Emergencies

Alex Ivanov

Published by Evergreen Publishing Pty Ltd, Australia

First published 2026

Published for Australia.

Available in digital formats

The information in this book is provided for educational and illustrative purposes only. It does not constitute medical, legal, engineering, or other professional advice, and should not be relied upon as such. Water sources, regulations, climate conditions, and infrastructure vary significantly by location and can change over time; information that is current at the time of writing may become outdated or may not apply to your specific circumstances. Nothing in this book constitutes a guarantee of safety, water potability, or survival outcomes in any emergency, wilderness, or disaster scenario. Results will vary based on individual circumstances, environment, equipment, physical condition, and execution, and no particular outcome is guaranteed. The techniques, equipment recommendations, and emergency procedures described in this book carry inherent risks, including the risk of illness, injury, or death, and readers assume all such risks in choosing to apply them. The author and publisher accept no liability for any loss, damage, illness, or injury arising from reliance on the contents of this book. Examples and scenarios included in this book are illustrative and, where based on real events, identifying details have been changed to protect the privacy of individuals involved. Always seek qualified professional advice, including from a medical practitioner, a licensed plumber or driller, or the relevant water authority, before making decisions or undertaking any water storage, sourcing, or treatment activity specific to your situation.

Published under pen name. The author's identity is withheld by mutual agreement with the publisher.

EVERGREEN
PUBLISHING
PTY LTD

PART I FOUNDATIONS

PART II HOME STORAGE

PART III SOURCING OFF-GRID

PART IV GREYWATER

CHAPTER 1

Why Water Comes First

The human body can survive for approximately three weeks without food, but only three days without water. This is the foundation of the "rule of threes" used throughout survival medicine: a person can survive roughly three minutes without breathable air, three hours without adequate shelter in extreme temperatures, three days without water, and three weeks without food. Shortages of food are unpleasant and, over a longer timeframe, can become dangerous. Shortages of water become dangerous far more quickly, and the margin for error is considerably smaller.

The three-day figure is an average calculated for a resting adult in moderate conditions, with no illness, no exertion and no pre-existing fluid deficit. It should be treated as an absolute outer limit, not a planning target. In Australian summer conditions, at 40 degrees Celsius with direct sun exposure and physical activity, meaningful physical and cognitive impairment from dehydration can begin within two to three hours. Serious incapacitation can follow within four to six hours in the most extreme heat conditions. A household relying on the standard three-day figure in a Queensland heatwave, with the added burden of a power outage and no functioning air conditioning, may have considerably less time than that figure suggests.

The general planning standard used by agencies including the United States Centers for Disease Control and Prevention (CDC) is a minimum of one gallon (approximately 3.8 litres) of water per person, per day, scaled upward for hot climates, pregnancy, illness, high activity levels and pets. This is discussed in full, with Australian-specific adjustments, in Chapter 2.

Water is not simply the item consumed to satisfy thirst. It is a required input for nearly every other function a household depends on during a disruption, including:

Medication preparation. Oral medications and, in particular, any reconstituted liquid medication require clean water for safe administration.

Infant formula preparation. Formula-fed infants cannot safely go without properly prepared water for more than a few hours before the combined risks of dehydration and improperly mixed formula become significant.

Wound cleaning and first aid. Disruptions produce a higher rate of cuts, burns and puncture injuries than ordinary daily life. Effective wound cleaning requires clean water; using contaminated water in its place is a direct pathway to infection.

Hand hygiene. Handwashing before food preparation and after using the toilet remains the single most effective barrier against the gastrointestinal illness that spreads quickly through households living at close quarters with reduced sanitation.

Food preparation. Rehydrating stored and dehydrated food, cooking staples such as rice and pasta, and washing fresh produce all draw on the same limited reserve.

Basic sanitation. Once mains water pressure fails, even a manually operated toilet requires a water source for flushing.

When a household's water supply is exhausted, it does not lose a single function. It loses the underlying resource that supports every other function at once, in a set of circumstances where every other challenge is already more demanding than normal. A household that runs out of water during a flood, a heatwave blackout or a supply chain disruption is not experiencing an isolated water problem. It is experiencing every other problem simultaneously, with reduced physical capacity, impaired judgement and an increased likelihood of the illnesses that make the remaining problems worse.

The Physiological Cost of Dehydration

Dehydration is measured as a percentage of body weight lost as fluid, and its effects follow a predictable and well-documented progression.

1 to 2 percent fluid loss. Thirst becomes noticeable. Measurable impairment begins in concentration, short-term memory, reaction time and mood, even before any physical symptoms are obvious to the affected person.

3 to 4 percent fluid loss. Physical performance declines noticeably. Headache, reduced appetite and dry mouth typically appear, along with a measurable reduction in physical work capacity.

5 to 6 percent fluid loss. Difficulty concentrating, irritability, sleepiness, and a rise in body temperature and heart rate all become apparent. Heat exhaustion becomes a realistic risk during any physical exertion.

7 to 10 percent fluid loss. Muscle spasms, dizziness, and a significant decline in physical and mental performance occur. This level of dehydration constitutes a medical emergency requiring urgent rehydration and, in most cases, professional medical attention.

Above 10 percent fluid loss. This level is life-threatening and can result in organ failure and death without prompt medical intervention.

Fluid loss accelerates sharply with heat and exertion. A sedentary adult at 40 degrees Celsius can lose approximately 500 millilitres of fluid per hour through perspiration alone. An active adult in the same conditions can lose more than a litre per hour. Children lose fluid more rapidly than adults relative to their body weight, and are often less reliable at recognising and reporting thirst. Elderly household members frequently have a blunted thirst response, meaning they can reach a meaningful level of dehydration before they feel thirsty at all.

This progression has a direct and practical implication for household preparedness planning. The decisions a household most needs to make well during a disruption, including whether to evacuate, how to triage an injury, and whether conditions are safe to remain in place, are frequently made under exactly the physical conditions that most degrade the judgement required to make them. A household plan that accounts for physical survival but not for this cognitive decline is an incomplete plan.

Beyond Drinking: The Functional Cost of Water Loss

The consequences of losing water access extend well beyond thirst, and are worth setting out in full, because most households underestimate the breadth of the problem until they experience it directly.

Hygiene deteriorates first, and illness follows closely behind. Without water for handwashing, gastrointestinal illness spreads rapidly through a household. This creates a compounding problem: a household that has already lost easy access to clean water is at greater risk of a gastrointestinal illness that itself accelerates fluid loss, through vomiting and diarrhoea, at precisely the point when fluid is hardest to replace.

Stored food becomes unusable. Dehydrated and freeze-dried emergency food, which forms a significant proportion of most household emergency pantries, is entirely dependent on water for preparation. A pantry stocked with rice, pasta, dried beans and dehydrated meals is not a usable food reserve if no water is available to prepare it.

Cooling options disappear. Wet cloths, sponge bathing and other forms of evaporative cooling are among the most effective available responses to heat stress, and all require a water supply that most households assume will always be present.

Vehicle and equipment maintenance becomes more difficult. Radiator top-ups, battery terminal cleaning and basic wound care while away from home all draw, in small quantities, on the same limited reserve.

Pets and animals become an immediate welfare responsibility. A dog in ordinary conditions requires approximately 250 millilitres of water per kilogram of body weight per day; this requirement rises sharply in heat. A 30-kilogram dog can require 7.5 litres or more per day in hot conditions. Households with poultry, horses or other animals carry a proportionally larger obligation that is frequently overlooked when preparedness planning focuses only on the human members of the household.

None of these outcomes are unusual or extreme. They are the ordinary, predictable consequence of a household losing access to the one resource that every other survival function depends on.

Case Studies in Water Failure

Understanding how water systems actually fail, rather than assuming they will simply continue working, is a useful foundation for the planning that follows in later chapters.

Cape Town's "Day Zero." Between 2017 and 2018, Cape Town, South Africa's second-largest city, came within months of officially exhausting its municipal water supply following several consecutive years of severe drought. City authorities prepared contingency plans to shut off household mains supply entirely and require residents to collect a strictly rationed daily allowance from approximately 200 designated collection points across the city. Day Zero was ultimately avoided through a combination of drastic mandated consumption reductions and late-arriving rainfall, but the episode remains one of the clearest modern demonstrations that a large, modern city can approach total water failure through drought alone. It illustrates why household-level water storage functions as insurance against exactly this kind of slow-developing municipal failure, regardless of how unlikely such a failure may seem in ordinary times.

Australian flood events. Queensland and New South Wales flood events have repeatedly demonstrated a different failure mode: not an absence of water, but an abundance of unsafe water. Mains water supply is frequently declared unsafe for consumption during and after major floods, because treatment plants are overwhelmed, damaged or inundated. Boil-water notices issued under these circumstances can remain in place for days or, in more serious cases, weeks. Households that assume running water from a tap is automatically safe water are poorly placed to respond to this kind of event. A stored reserve of water of known, verified quality is arguably more valuable during a flood than during almost any other type of disruption, precisely because the tap continues to run while supplying water that may not be safe to drink.

Extended power outages. Queensland's history includes prolonged, infrastructure-driven power disruptions, and more recent severe weather events have repeatedly shown how quickly a power outage becomes a water crisis. Electric water pumps, water-dependent cooling systems, and, in properties supplied by rainwater tank or bore with an electric pressure pump, the entire household water supply can fail simultaneously once the power fails. A household on tank water with an electric pump and no manual override may have several thousand litres of water in storage that it is physically unable to access.

A Framework for Preparedness: Mild, Moderate and Extreme

Different types of disruption require different levels of water preparedness. It is useful to plan across three tiers, rather than treating household preparedness as a single, fixed state.

Mild disruptions. These include a burst water main, a short-term boil-water notice following heavy rainfall, or a blackout lasting 24 to 48 hours. A household with a properly stocked 72-hour water kit, covered in Chapter 3, experiences minimal disruption from events of this scale.

Moderate disruptions. These include a significant flood, the aftermath of a cyclone, or a multi-week disruption to the supply of bottled water and water treatment products. A two-week water reserve, supplemented by rainwater harvesting and, where appropriate, a household greywater system, provides an adequate response to disruptions of this scale.

Extreme disruptions. These involve prolonged failure of mains water infrastructure over a period of months, requiring a genuine capacity for self-sufficiency through wells, bores, large-scale rainwater capture, and full-spectrum water treatment. Relatively few Australian households need to plan seriously for this tier, but those that do require the material in this book covered in real depth rather than summarised briefly.

Most Australian households will find that solid preparation for the first two tiers covers the overwhelming majority of realistic scenarios. This book is weighted accordingly. However, the material required to plan for extended disruption is included in full, because the practical difference in cost and effort between preparing for moderate disruption and preparing for extended disruption is often smaller than it first appears; frequently, it is the difference between installing one rainwater tank and installing two, or between owning one water filter and building a small, complete filtration and treatment system capable of supporting a household indefinitely.

How This Book Is Organised

Every chapter that follows works toward a single practical outcome: a household that can state, specifically and honestly, how many days of clean water it has access to at any given time, and that understands exactly how to extend that figure when circumstances require it.

The material is organised in the sequence a genuine preparedness build tends to follow in practice.

Part II addresses home storage first, because stored water is the fastest, most affordable and least skill-dependent layer of water security available, and should be established before any more advanced system is considered.

Part III addresses sourcing water directly from the environment once stored supplies are exhausted, including reading the landscape, harvesting dew and precipitation, extracting water from plants, digging for water, and locating water in places most households never think to look.

Part IV addresses greywater recycling, one of the most underused resources already present in most Australian homes.

Parts V and VI address testing and treatment in genuine depth, because locating a water source and confirming that it is safe to drink are two entirely separate achievements. Treating them as equivalent is a common and dangerous error.

Part VII brings the household material together, covering the needs of vulnerable household members, the recognition and treatment of water-related illness, water rationing and conservation under scarcity, and the construction of a complete, actionable household water security plan.

Part VIII covers water strategy for extreme and low-probability scenarios: isolation in remote wilderness or at sea, infrastructure destroyed by war, sabotage, cyberattack or supply chain collapse, natural disasters severe enough to destroy water systems outright, and entrapment or mass displacement. This part is written for the reader who wants the outer edge of preparedness covered as thoroughly as the everyday case.

The great majority of this book, Parts II through VII, requires no rural acreage, no off-grid property and no substantial one-off expense. It requires only that water be treated as this chapter has argued it should be treated throughout the remainder of the book: as the first and most urgent item in any preparedness plan, rather than an afterthought positioned below torches, batteries and tinned food.

Part VIII departs from that standard for readers who want it. It sets out water strategy for a set of extreme and low-probability circumstances, including infrastructure destroyed by war or disaster, isolation in remote wilderness or at sea, and entrapment or mass displacement, where the ordinary household approach no longer applies and a different, more extreme skill set is required.

Sources referenced in this chapter

Centers for Disease Control and Prevention (CDC), How to Create and Store an Emergency Water Supply, cdc.gov/water-emergency

PetMD and Cornell University Feline Health Center, companion animal hydration guidance

Australian Government, Bureau of Meteorology, climate and heatwave data

Historical and news reporting on the Cape Town "Day Zero" water crisis, 2017 to 2018

CHAPTER 2

Calculating Your Household's Real Water Requirement

Most published water storage advice begins with a single number, one gallon, or approximately 3.8 litres, per person, per day. This figure originates from United States federal emergency planning guidance and is repeated across the majority of preparedness literature without adjustment (CDC). It is a reasonable minimum for a resting adult in a temperate climate consuming water for drinking, basic food preparation and minimal hygiene. It is not an adequate planning figure for an Australian household, and treating it as one is one of the most common and consequential errors in household water preparedness.

This chapter sets out how to calculate a household-specific figure that accounts for climate, activity, household composition, medical circumstances and animals, and how to convert that figure into the specific volumes required for the storage tiers covered in Part II.

Why the Baseline Figure Understates Australian Need

Three factors specific to Australian conditions push real household water requirements well above the international baseline.

Heat. Large parts of Australia regularly exceed 35 degrees Celsius for extended periods, and capital cities including Brisbane, Sydney, Adelaide and Perth experience multiple heatwave events most summers. Fluid loss through perspiration increases sharply with temperature, and a planning figure calculated for a temperate climate will consistently underestimate real consumption during exactly the conditions in which an emergency is most likely to occur.

Distance and delay. Many Australian households, including a significant proportion of regional and rural properties, are a considerable distance from the nearest alternative water source or emergency distribution point. A planning figure that assumes resupply within three days is unsuitable for a property that may be isolated by flood, fire or road closure for a week or more.

Housing stock. Older Australian housing, and a meaningful proportion of newer housing, is not built to the thermal performance standards common in some other developed countries. Without functioning air conditioning during a blackout, indoor temperatures in an Australian home can climb well above outdoor ambient temperature, increasing fluid loss for every household member inside it.

For these reasons, this book uses four litres per person, per day, as the baseline planning figure for a household in normal Australian conditions undertaking light activity, rather than the 3.8-litre international minimum. This figure is increased further, as set out below, for heat, activity, pregnancy, illness, age and pets.

The Component Parts of Daily Water Need

Total daily water requirement is the sum of several distinct uses, each of which should be considered individually rather than folded into a single vague estimate.

Drinking water. The largest single component. A sedentary adult in moderate conditions requires approximately two litres per day. This rises to three litres or more in hot conditions, and can exceed four litres per day during heavy physical activity in extreme heat.

Food preparation. Cooking rice, pasta, dried beans and rehydrating dehydrated meals typically requires between 0.5 and one litre per person, per day, depending on the specific foods stored.

Basic hygiene. Handwashing, teeth cleaning and minimal sponge bathing require approximately 0.5 to one litre per person, per day at a bare minimum, and considerably more if any showering is possible.

Medication and medical needs. A smaller quantity, typically well under 0.5 litres per day for most households, but non-negotiable and non-substitutable for any household member on regular medication.

Summed together, these components produce a figure in the range of four to six litres per person, per day for an average adult household in normal Australian conditions engaged in light activity. This is the figure that should be used for the two-week reserve calculations in Chapter 4, adjusted upward where any of the factors below apply.

Adjusting for Heat and Activity

Heat and physical exertion are the two largest variables affecting real water need, and both should be adjusted for explicitly rather than absorbed into a generic buffer.

Heat adjustment. For any period in which daytime temperatures are expected to exceed 35 degrees Celsius, particularly without functioning air conditioning, increase the baseline drinking water allowance by 50 to 100 percent. A household planning four litres per person per day in moderate conditions should plan for six to eight litres per person per day during a heatwave.

Activity adjustment. Any household member engaged in sustained physical activity, including clearing storm or flood debris, carrying supplies over distance, or manual repair work, should be allocated an additional 0.5 to one litre per hour of activity, on top of the baseline resting requirement.

Combined heat and activity. When both factors apply simultaneously, for example clearing debris in direct sun during a heatwave, the two adjustments should be applied together rather than treating the larger of the two as sufficient. This can push individual daily requirement above eight litres for a single household member on the most demanding days.

Adjusting for Household Composition

A single household-wide figure, multiplied by the number of people in the home, is a reasonable starting point, but several categories of household member require an individual adjustment.

Infants and young children. Children require less total volume than adults in absolute terms, but considerably more relative to body weight, and are at higher risk from both dehydration and unsafe water because of less mature immune systems and kidney function. Plan for a minimum of one litre per day for an infant, rising with age, and treat any water intended for formula preparation or direct consumption by a young child as requiring the highest standard of treatment available in the household, regardless of the treatment applied to water for adult use.

Pregnant and breastfeeding women. Fluid requirements increase measurably during pregnancy and increase further during breastfeeding, when the body is producing milk with a high water content. An additional one litre per day above the standard adult allowance is a reasonable planning adjustment for both.

Elderly household members. Older adults frequently have a blunted thirst response and, in some cases, medical conditions or medications that affect fluid balance. Rather than relying on thirst as an indicator, plan a fixed daily allowance for elderly household members and monitor actual intake, particularly during heat events.

Household members with medical conditions. Diabetes,

kidney disease, and a range of other chronic conditions can significantly increase fluid requirements or introduce specific water quality requirements, such as the dialysis and immunocompromised considerations covered in Chapter 20. Any household with a member in this category should seek specific guidance from their treating clinician as part of building the household water plan, rather than relying solely on general planning figures.

Adjusting for Pets and Animals

Animal water requirements are frequently omitted from household planning entirely, despite representing a meaningful volume in many Australian households.

Dogs. Approximately 250 millilitres per kilogram of body weight per day in moderate conditions, increasing significantly in heat or with activity. A 30-kilogram dog can require 7.5 litres or more per day in hot conditions.

Cats. Approximately 200 to 250 millilitres total per day for an average adult cat in moderate conditions, though cats obtain a portion of their fluid intake from food, and this figure should be treated as an approximate minimum rather than a precise requirement.

Poultry. A single chicken typically requires approximately 200 to 500 millilitres per day, with requirements increasing in hot weather; a household keeping backyard chickens should plan for this as a fixed daily draw on stored supply.

Livestock and larger animals. Horses, sheep, cattle and other larger animals have substantially higher daily water requirements, frequently in the tens of litres per animal per day, and households keeping any of these animals should treat livestock water supply as a distinct planning category with its own storage and sourcing strategy, rather than an extension of the household's drinking water reserve.

Worked Examples:

The following worked examples illustrate how these adjustments combine in practice, planning for a two-week reserve in normal Australian conditions with the possibility of a heatwave during the disruption period.

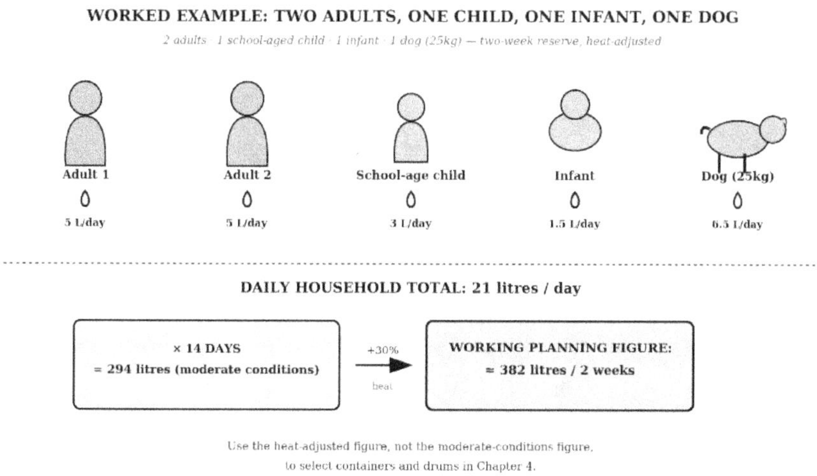

Adult 1 (moderate activity): 5 litres per day

Adult 2 (moderate activity): 5 litres per day

School-aged child: 3 litres per day

Infant: 1.5 litres per day (including formula preparation)

Dog (25 kilograms): 6.5 litres per day

Daily household total: 21 litres per day

Two-week total (14 days): 382 litres

Two-week total in heat (14 days +30%): 355 litres

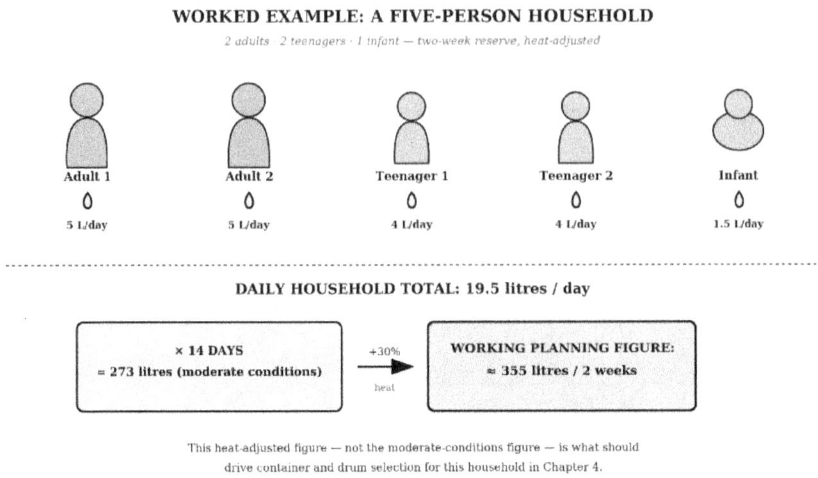

Adult 1 (moderate activity): 5 litres per day

Adult 2 (moderate activity): 5 litres per day

Teenager 1: 4 litres per day

Teenager 2: 4 litres per day

Infant: 1.5 litres per day (including formula preparation)

Daily household total: 19.5 litres per day

Two-week total (14 days): 273 litres

Two-week total in heat (14 days +30%): 355 litres

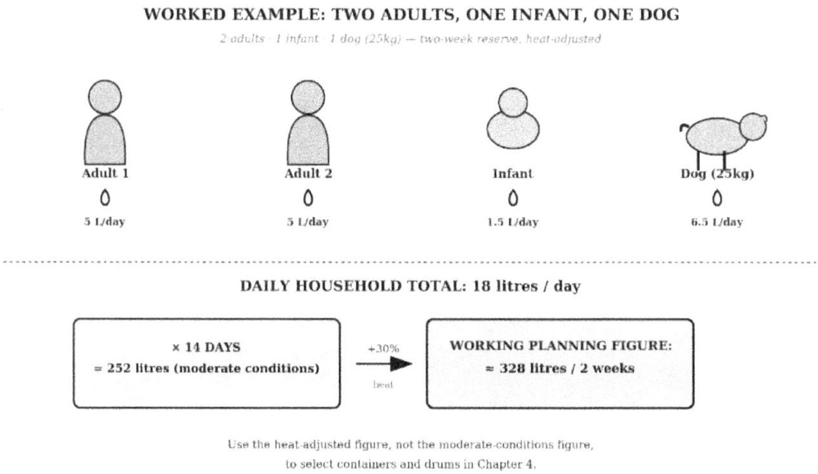

Adult 1 (moderate activity): 5 litres per day

Adult 2 (moderate activity): 5 litres per day

Infant: 1.5 litres per day (including formula preparation)

Dog (25 kilograms): 6.5 litres per day

Daily household total: 18 litres per day

Two-week total (14 days): 252 litres

Two-week total in heat (14 days +30%): 328 litres

Applying a heat adjustment of approximately 30 percent to account for the realistic possibility of hot weather during the disruption period brings the working planning figure to approximately 380 litres for the two-week period. This is the figure this household should use when selecting storage containers and drums in Chapter 4, rather than the lower figure calculated for moderate conditions alone.

Building Your Own Figure

Readers are encouraged to complete the same calculation for their own household before proceeding to Part II, using the following steps.

List every household member individually, including any infants, and assign each an age-appropriate baseline drinking water figure.

List every pet and animal in the household and assign each an appropriate daily water figure based on species and body weight.

Add a food preparation and hygiene allowance of approximately one litre per person, per day, to each human household member's figure.

Sum all figures to produce a total daily household requirement in normal conditions.

Apply a heat adjustment of 30 to 100 percent, depending on the climate of the region and the likely season of any disruption, to produce a working daily figure.

Multiply the working daily figure by the number of days being planned for (three days, fourteen days, or longer, as covered in Parts II and VIII) to produce the target storage volume for that planning tier.

This figure, calculated specifically for your household rather than borrowed from generic guidance, is the number that should drive every storage and equipment decision made in the remainder of this book.

Sources referenced in this chapter

Centers for Disease Control and Prevention (CDC), How to Create and Store an Emergency Water Supply, cdc.gov/water-emergency

Ready.gov, Water, ready.gov/water

PetMD and Cornell University Feline Health Center, companion animal hydration guidance

Australian Bureau of Meteorology, heatwave and climate data

CHAPTER 3

The First 72 Hours

The first 72 hours after any disruption are the period in which a household is most exposed and least equipped to improvise. Distribution points have not yet been established, mains supply may or may not have failed outright, and the household is typically still assessing the scale of the problem rather than actively solving it. A properly stocked 72-hour water supply removes water entirely from the list of problems that need solving during this window, allowing attention to go to shelter, safety and information instead.

Using the four-litre-per-person, per-day baseline established in Chapter 2, a household of four adults requires a minimum of 48 litres for 72 hours. This figure should be adjusted upward for heat, activity, pregnancy, illness, elderly household members and pets exactly as set out in Chapter 2. A household with two adults, two children and a dog, planning for a realistic Australian summer scenario, should plan for closer to 60 litres for the same three-day period once these adjustments are applied.

Selecting the Right Container for 72-Hour Storage

Three container categories suit 72-hour storage, and each has a distinct role.

Commercial bottled water. Sold in every supermarket in Australia in sizes from 600 millilitres to five litres, commercial bottled water is sealed, food-grade, carries a printed use-by or bottling date, and requires no preparation before storage. Its main disadvantages at scale are cost per litre, storage volume inefficiency relative to larger containers, and plastic waste. For 72-hour planning specifically, these disadvantages matter less than the convenience, and commercial bottled water remains a reasonable primary or supplementary option at this scale.

Food-grade jerry cans. Polyethylene jerry cans in 10 to 20-litre sizes are widely available from camping retailers, hardware stores and some supermarkets, and represent better value than bottled water once a household moves beyond the smallest emergency kit. When selecting a jerry can for water storage, confirm the following:

The container is explicitly labelled food-grade, or carries a food-safe triangle symbol on the base.

The lid seals completely, with no leakage when the filled container is tilted or laid on its side.

The handle is robust enough to carry the full container, which will weigh approximately one kilogram per litre, over a meaningful distance.

The container is opaque or UV-resistant, to reduce the risk of algal growth in stored water exposed to light.

Repurposed containers, used only as a last resort. Clean, previously unused two-litre soft drink bottles can be used if no food-grade option is available, but containers that previously held milk, fruit juice or any chemical product should never be used for water storage. Milk proteins and fruit sugars cannot be reliably removed from plastic, and residual chemical contamination from cleaning products, fuel or pesticides poses a genuine health risk regardless of how thoroughly the container appears to have been rinsed.

Filling, Labelling and Storing

Correct preparation of a 72-hour water supply is straightforward, but skipping any of the following steps introduces avoidable risk.

Clean before filling. Wash any reused container with dishwashing detergent and rinse thoroughly. For containers with an uncertain prior use, sanitise with a diluted, unscented household bleach solution before use.

Fill from a trusted source. Ordinary treated mains tap water is appropriate for 72-hour storage in the great majority of Australian households and does not require additional treatment before storage.

Label with the fill date. Use a permanent marker directly on the container, not on a label that may detach or become illegible. This single step is the difference between confidently using stored water and being uncertain of its age during an actual emergency.

Store in a stable, moderate-temperature location. A cool interior room, or a shaded section of a garage that does not reach extreme summer temperatures, is appropriate. A car boot in an Australian summer is not, and water stored under these conditions should be treated as compromised and replaced.

Keep away from chemicals. Store water separately from cleaning products, fuel, pesticides and other chemicals, since even a well-sealed container can absorb odours and, in the case of a leak or spill nearby, actual contamination.

Rotation and Testing

Commercial bottled water should be rotated according to the manufacturer's printed date. Water filled at home in food-grade containers does not spoil in the way food does, but taste can deteriorate over time, particularly in plastic containers exposed to any warmth, and the practical guidance from agencies including the CDC is to replace home-filled water approximately every six months as a matter of routine (CDC).

Testing a 72-hour kit is a physical task, not a paperwork exercise. Every three to six months:

Confirm every container is in its designated storage location and has not been repurposed for another use.

Check that lids are intact and sealing properly, with no visible warping or cracking.

Confirm fill dates are within the intended rotation period and replace any water that has exceeded it.

Open one container at random and check for any unusual smell, cloudiness or discolouration, replacing the batch if anything is observed.

The Bathtub Pre-Fill: The Single Most Effective Free Action Available

When a household has advance warning of an imminent disruption, whether a flood warning, a severe storm warning, or advice of an imminent, planned mains water interruption, the single most effective water action available at essentially zero cost is filling every bathtub and every clean, large container in the home immediately.

This action takes approximately ten minutes in a household with two bathrooms and produces 300 to 400 litres of utility water at no cost beyond the time required. A standard Australian bathtub holds approximately 150 to 200 litres. This water is not immediately drinking-quality without treatment, because bathtub surfaces are not sterile and the water will pick up surface residue on contact, but it is immediately usable for several critical purposes:

Toilet flushing, by pouring directly into the bowl, which requires no functioning mains water pressure.

Cleaning of surfaces, hands and equipment.

Cooling for heat-affected household members, through wet cloths or sponge bathing.

Cooking and drinking, once treated by boiling or chemical disinfection as set out in Part VI.

The timing of this action is critical, and it is the single most common mistake households make. Filling must occur before the disruption reaches the property, not after. A flood warning issued several hours ahead provides the necessary window. A severe storm warning issued for the afternoon provides the necessary window. Once mains water pressure has already dropped or failed, the opportunity has passed. In Queensland and other flood-prone regions of Australia, filling all available bathtubs and containers should be treated as one of the first five actions taken on receipt of a flood warning that carries any risk of water supply contamination or interruption.

When There Is No Warning

Not every disruption provides this window. A burst water main, a contamination incident forcing an immediate shutdown, or the sudden failure that can follow an earthquake or a serious accident at a treatment plant can remove mains water without any warning at all, and the bathtub pre-fill strategy described above is only ever available to a household that had advance notice. This is precisely why the pre-fill is described as the single most effective free action available, rather than the foundation of 72-hour planning: it is a valuable bonus when circumstances allow it, not something the household's actual water security should depend on. A 72-hour kit built and stored in advance, to the figure calculated in Chapter 2, must be sufficient on its own, with or without any opportunity to fill a bathtub first.

When mains water fails without warning, the household's first action should be to capture whatever water remains in the system rather than to begin searching for new sources. Taps at the lowest point in the house, opened immediately, will often release a further volume of water already sitting in the pipes before air enters the line, and the hot water system, covered in full in Chapter 11, typically holds a further 125 to 315 litres depending on its size, most of which remains usable once the unit is isolated from the mains supply. Once this residual water has been captured and added to the stored 72-hour kit, the household should move directly into the rationing discipline set out in Chapter 22, rather than assuming normal use can continue until the true scale and likely duration of the disruption becomes clear.

Purpose-Built Bathtub Bladders

A category of product exists specifically to make the bathtub pre-fill strategy more effective and more hygienic: large food-grade plastic bladders designed to fill a standard bathtub completely from the tap, isolating the stored water from direct contact with the bathtub surface. The WaterBOB is the product most households will encounter in this category and has become the de facto standard, though equivalent products from other manufacturers work on the same principle.

These bladders hold up to approximately 450 litres per unit, fold flat for storage when empty, and cost in the order of thirty to fifty Australian dollars from camping and emergency preparedness retailers. Because the water is contained within a sealed food-grade membrane rather than sitting directly in the bathtub, it remains at a meaningfully higher standard of cleanliness than an open-filled bath, and is generally suitable for drinking after a lower level of treatment than open bathtub water would require.

A household with two of these bladders in storage has approximately 900 litres of protected emergency water capacity that can be deployed in the time it takes to unfold the bladder, position it in the bath and connect it to the tap. For households in flood-prone areas, or any area with a known history of water supply interruption, keeping two of these bladders in storage alongside standard water containers is a reasonable and inexpensive addition to a 72-hour and two-week preparedness plan alike.

Building the 72-Hour Kit: A Practical Checklist

- Calculate the household's specific 72-hour requirement using the method in Chapter 2, including all pets and any heat or activity adjustment.
- Select a combination of commercial bottled water and food-grade jerry cans that meets this figure, favouring jerry cans for the majority of the volume once the household moves beyond the smallest kit size.
- Store the kit in a single, known, accessible location that every adult in the household can locate without assistance, including in darkness.
- Label every container with the fill date and set a recurring reminder to review the kit every three to six months.
- Keep at least one purpose-built bathtub bladder in storage if the household lives in a flood-prone area or an area with any history of water supply interruption.
- Confirm every household member, including older children, knows where the kit is stored and understands that stored water is for use during a genuine disruption, not everyday convenience.

The 72-hour kit is the foundation the remainder of this book builds on. Chapter 4 extends this same approach to the considerably larger volumes required for a genuine two-week household reserve.

Sources referenced in this chapter

Centers for Disease Control and Prevention (CDC), How to Create and Store an Emergency Water Supply, cdc.gov/water-emergency

Ready.gov, Build a Kit, ready.gov/kit

Ready.gov, Water, ready.gov/water

Queensland Government, flood preparedness and household warning guidance

CHAPTER 4

The Two-Week Reserve

A 72-hour kit answers the first, most urgent question a household faces during a disruption. A two-week reserve answers the more consequential question that follows it: what happens if the disruption does not resolve within three days. Australian experience across floods, cyclones and extended power outages shows that a genuinely disruptive event affecting water supply frequently runs well beyond the 72-hour mark, and the CDC's own guidance, while using a three-day figure as its baseline recommendation, explicitly encourages households to build toward a two-week supply where practical.[^cdc2]

This chapter sets out how to store the considerably larger volume a two-week reserve requires, without the cost or the storage footprint becoming unmanageable.

The Volume Involved

Using the four-litre-per-person, per-day baseline from Chapter 2, a household of four adults requires a minimum of 224 litres for a two-week period. This is a genuine volume that requires deliberate planning for container selection, storage location and rotation, rather than an incidental addition to the pantry. It is, however, entirely achievable in a typical Australian home without structural modification, and represents a one-time purchase that, with routine maintenance, lasts for years.

For the worked example household from Chapter 2, two adults, one school-aged child, one infant and a medium-sized dog, the adjusted two-week figure came to approximately 380 litres once a heat adjustment was applied. This chapter uses figures in the 220 to 400 litre range as the working planning target for a typical household, and readers should substitute their own calculated figure from Chapter 2.

Large-Volume Container Options

Three container categories dominate two-week and larger household storage, and each suits a different combination of budget, available space and priority.

Large food-grade polyethylene drums, 100 to 200 litres. Available from camping and emergency preparedness retailers, agricultural suppliers and some hardware stores, these drums are purpose-built for water storage, with food-grade interior surfaces, UV-stabilised exteriors, and a fitted tap or bung for controlled dispensing. Two 110-litre drums provide the planning minimum for a four-person household and occupy a floor footprint roughly equivalent to a standard refrigerator, making them highly space-efficient for the volume stored. A filled 110-litre drum, sealed and stored in a cool location out of direct sunlight, maintains acceptable water quality for two to five years before treatment or replacement is required.

Food-grade jerry cans, 20-litre sizes. These complement large drums by providing portability that a drum cannot offer once full. A full 110-litre drum weighs 110 kilograms and cannot realistically be moved by most adults. A full 20-litre jerry can weighs approximately 20 kilograms and remains manageable for most adults in reasonable health, allowing selective portability during an evacuation without requiring the household to abandon its full stored reserve.

Intermediate bulk containers (IBC totes), 500 to 1,000 litres. For households with adequate space, typically in a garage, shed or under-house area, an IBC tote provides very large-volume storage in a single unit. These are more commonly sourced second-hand from agricultural or industrial suppliers and require careful verification that the tote's prior contents were food-safe before repurposing for water storage. An IBC tote of unknown or non-food history should never be used for drinking water storage without thorough cleaning and, ideally, confirmation of its prior use from the supplier.

Container Placement and Structural Considerations

Water is heavy. A single litre weighs one kilogram, meaning a filled 200-litre drum weighs 200 kilograms and an IBC tote at full capacity can weigh close to a tonne. This has practical implications that are easy to overlook when focused solely on the container itself.

Floor loading. Concrete garage floors and ground-level slabs generally tolerate this weight without issue. Elevated timber floors, particularly in older Queenslander-style homes on stumps, should be assessed before placing large drums or totes, since concentrated point loads of this scale were not necessarily accounted for in the original structural design.

Level, stable ground. Large containers should sit on a level surface. An uneven base can place uneven stress on the container wall over time and increase the risk of a slow leak developing at a stress point.

Accessibility. Position large containers so the tap or bung is accessible without needing to move the container itself, and leave enough clearance around the unit to attach a hose or fill smaller containers directly.

Away from direct sun and heat sources. Direct sunlight accelerates plastic degradation and promotes algal growth even in nominally sealed containers. A shaded section of a garage, a laundry, or an under-house area in a home built on stumps are all suitable locations in most Australian climates.

Filling and Maintaining Bulk Storage

The process for filling large containers follows the same principles as the smaller containers covered in Chapter 3, scaled up.

Rinse the interior of the drum or tote with a small quantity of clean water before the first fill, even if the container is new, to clear any manufacturing residue or dust.

Fill from the mains tap using a food-grade hose, never a hose that has been used for chemicals, fuel or garden treatments.

Seal the container completely once filled, and label it clearly with the fill date using a permanent marker on the container itself.

Position the container in its permanent storage location before it is completely full where possible, since moving a large, heavy container after filling is both difficult and a genuine injury risk.

Inspect quarterly for any sign of leaking, bulging or discolouration of the container wall, and inspect the tap or bung for proper sealing.

A water preservative treatment, available from emergency preparedness suppliers, can extend the viable storage life of bulk-stored water toward the upper end of the two-to-five-year range, and is a reasonable addition for households storing water for extended periods without planning to rotate it on a strict schedule. Where a preservative product is used, follow the manufacturer's dosing instructions exactly, since both under-dosing and over-dosing reduce the product's effectiveness.

Rotation Strategy at Scale

Rotating 220 litres of water is a more deliberate undertaking than rotating a 72-hour kit, and a workable household system should be established from the outset rather than improvised later.

Stagger fill dates where practical. A household with two 110-litre drums can fill and label them at different times, rather than both simultaneously, so that rotation and testing can be spread across the year rather than requiring a single large event.

Use the oldest water first for non-drinking purposes. Water approaching the end of its planned rotation period can be diverted to garden irrigation, cleaning or car washing rather than discarded, with the container then refilled and relabelled.

Set a fixed annual date. Choosing a specific, memorable date each year, such as the start of the Australian storm season, to inspect and, where necessary, refresh the bulk reserve creates a habit that is far more reliable than an informal intention to "check on it sometime."

Combining the Two-Week Reserve With the Bathtub Strategy

The bathtub pre-fill strategy introduced in Chapter 3 adds meaningfully to a two-week plan at zero additional equipment cost. When a major flood warning, severe storm warning or planned water supply interruption is clearly imminent with several hours of notice, filling every bathtub and every additional clean container in the home can add 300 to 400 litres of utility water in approximately ten minutes. Combined with a properly stocked two-week reserve of stored drinking water, this bathtub water becomes the household's primary supply for non-potable uses such as toilet flushing and cleaning, preserving the higher-quality stored reserve specifically for drinking, food preparation and medical needs for longer.

A Practical Two-Week Storage Plan

- Calculate the household's specific two-week figure using the method in Chapter 2, including all relevant adjustments.

- Select two or more 100 to 200-litre food-grade drums as the primary storage method for most households, supplemented by 20-litre jerry cans for portability.

- Position drums on a stable, level, shaded surface with adequate floor support, away from direct sunlight and any chemical storage.

- Label every container with its fill date and establish a fixed annual date for inspection and rotation.

- Consider a water preservative treatment for households that do not intend to rotate stored water on a strict schedule.

- Treat the two-week reserve as distinct from, and in addition to, the 72-hour kit covered in Chapter 3, rather than assuming one substitutes for the other.

Sources referenced in this chapter

Centers for Disease Control and Prevention (CDC), How to Create and Store an Emergency Water Supply, cdc.gov/water-emergency

Queensland Government, household structural and storm season preparedness guidance

Australian emergency preparedness retailer product specifications for food-grade water drums and bathtub bladders

CDC, How to Create and Store an Emergency Water Supply (cdc.gov/water-emergency)

CHAPTER 5

Long-Term Bulk Storage

Chapters 3 and 4 covered water intended for use within days to weeks of storage. This chapter addresses a different planning horizon: water intended to sit, largely untouched, for a year or more, as the base layer of a household's preparedness rather than an active, frequently rotated supply. Long-term bulk storage suits households planning explicitly for the "moderate" and "extreme" tiers set out in Chapter 1, and households in regional or remote locations where resupply after a disruption may be genuinely delayed rather than merely inconvenient.

Water Does Not Expire, But It Does Degrade

A common misconception is that stored water has a fixed expiry date in the way packaged food does. It does not. Pure water itself does not spoil. What changes over extended storage is not the water but its surroundings, and the practical risks that follow are real even though the underlying chemistry is different from food spoilage.

Container leaching. Even food-grade plastic can leach trace compounds into stored water over long periods, particularly in containers exposed to heat, affecting taste and, in poor-quality or non-food-grade containers, potentially safety.

Light exposure. Ultraviolet light penetrating a translucent or clear container promotes the growth of algae and other organisms if any organic material or airborne contamination entered the water during filling. This is why opaque, UV-resistant containers are specified throughout this book rather than clear ones.

Seal integrity. A container that seals well on day one may develop a slow leak in its lid gasket or bung over months or years of storage, allowing airborne contamination to enter gradually.

Initial contamination. Water that was not properly treated or that was filled into an improperly cleaned container carries whatever biological or chemical contamination was present at the time of filling, and that contamination does not necessarily resolve itself simply because time has passed.

The practical implication is that long-term storage is less about the water "going off" and more about maintaining the integrity of the system it sits inside: the container, the seal, the storage environment and, where used, the chemical treatment added at the time of filling.

Chemical Stabilisation for Extended Storage

For water intended to sit for a year or more without rotation, adding a stabilising treatment at the time of filling meaningfully extends the period over which the water remains safe to drink without further treatment.

Unscented household chlorine bleach, containing 5 to 9 percent sodium hypochlorite as the only active ingredient, is the standard, low-cost stabilisation method recommended by agencies including the CDC and Ready.gov.[^bleach] The standard dosing rate is:

- Confirm the bleach is unscented, contains no additional cleaning agents, and lists sodium hypochlorite in the 5 to 9 percent range as its only active ingredient. Scented bleaches, colour-safe bleaches and bleaches with added surfactants should never be used for water treatment.

- Add approximately one-eighth of a teaspoon of bleach per gallon (approximately 3.8 litres) of water, or scale proportionally for larger containers.

- Stir or agitate the container to distribute the bleach evenly, then allow the water to stand for at least 30 minutes before sealing for long-term storage.

- Label the container clearly with both the fill date and a note that the water has been chemically stabilised, since this affects how the water should be assessed before use, as covered later in this chapter.

Commercial water preserver concentrates, available from emergency preparedness suppliers, provide an alternative to bleach dosing and are formulated specifically for multi-year storage. These typically use a measured dropper or pre-portioned dose per standard container size, removing the calculation step required with household bleach, and are a reasonable choice for households that prefer a purpose-built product over calculating a bleach dose accurately at scale.

Long-Life Water Pouches

A distinct product category exists specifically for long-term storage: sealed, foil-lined water pouches, typically sold in cases of small individual pouches, with a manufacturer-rated shelf life of five years or more. These are considerably more expensive per litre than water stored in a household-filled drum, but offer genuine advantages for a specific role within a long-term storage plan.

They require no preparation, dosing or rotation planning by the household. The manufacturer has already solved the stabilisation problem.

They are individually sealed, so a failure or contamination event in one pouch does not affect the rest of the supply, unlike a single large drum where a contamination event affects the entire volume at once.

They are compact and suitable for vehicle kits, workplace kits, or any location where a large drum is impractical.

For most households, these pouches are best used as a supplementary layer, particularly for vehicle-based and portable kits, rather than as the primary method of meeting the full two-week or long-term volume calculated in Chapters 2 and 4, where the cost per litre makes drums and IBC totes considerably more economical for the bulk of the required volume.

Fifty-Five-Gallon Drums and Larger Bulk Systems

The 55-gallon (approximately 200-litre) drum is a long-standing standard size in bulk water storage, sitting between the 100 to 200-litre drums discussed in Chapter 4 and the largest IBC totes. These drums, along with any bulk storage system in this size range, benefit from several long-term-specific practices beyond the placement guidance already covered in Chapter 4.

Elevated tank stands. Raising a bulk storage drum or tank on a purpose-built stand allows gravity-fed dispensing without a pump, which matters considerably during a power outage. A stand needs to be rated for the full filled weight of the container, not the empty weight, and should be checked for stability before the container is filled rather than after.

Desludging. Sediment settles at the base of any large container over extended storage, even with clean source water and a stabilising treatment. Bulk containers with a base-mounted tap should be dosed and drawn from a slightly elevated outlet where possible, so sediment is not the first material drawn when the container is tapped, and the container should be fully drained and inspected for sediment build-up every two to three years.

Seasonal temperature cycling. Bulk containers stored in a garage or shed experience the full range of Australian seasonal temperature variation. Repeated expansion and contraction over years of storage is a more significant stress on seals and container walls than any single hot day, and is the primary reason for the periodic visual inspection schedule set out in Chapter 4.

Assessing Long-Stored Water Before Use

Water that has been correctly stabilised and stored for an extended period should still be assessed before use, rather than assumed safe purely on the basis of the stabilisation step taken years earlier.

- Check the container visually for any sign of leaking, bulging, or damage to the seal or bung.

- Open the container and check for any unusual smell. Correctly stabilised water may have a faint chlorine odour, which is expected and not a concern; a musty, sulphurous or otherwise off odour indicates the water should not be consumed without treatment.

- Check for visible cloudiness, discolouration or particulate matter, which may indicate sediment disturbance, container degradation or biological growth.

- Where available, use a simple chlorine test strip to confirm a residual level consistent with the original stabilisation dose. A residual well below the expected level suggests the stabilising treatment has broken down and the water should be re-treated, using the methods in Part VI, before use.

Water that fails any of these checks should not be assumed unusable outright. In most cases it can be brought back to a safe standard through the boiling, filtration or chemical treatment methods covered in Part VI, rather than discarded. Long-term storage is intended to reduce the amount of treatment required at the point of use, not to eliminate the need for a final safety check entirely.

Matching Long-Term Storage to Household Risk

The right amount to invest in long-term bulk storage depends on the household's realistic risk profile, using the mild, moderate and extreme framework introduced in Chapter 1.

Households primarily concerned with mild and moderate disruption are generally well served by the 72-hour and two-week systems covered in Chapters 3 and 4, with long-term storage limited to a modest additional buffer, such as one extra stabilised drum, rather than a large-scale system.

Households in regional or remote areas, or those explicitly planning for extended disruption, benefit from a genuine long-term bulk system: multiple stabilised 200-litre drums or an IBC tote, sized to the household's calculated requirement from Chapter 2 multiplied out to 30 days or more, supplemented by long-life pouches for portable and vehicle use.

Households planning for the extreme scenarios covered in Part VIII should treat bulk storage as one layer among several, working alongside the rainwater harvesting, well and bore, and field treatment capability covered in the following chapters, rather than attempting to store their entire theoretical requirement as static bulk water.

Sources referenced in this chapter

Centers for Disease Control and Prevention (CDC), How to Create and Store an Emergency Water Supply, cdc.gov/water-emergency

Ready.gov, Water, ready.gov/water

CHAPTER 6

Rainwater Harvesting Systems

Static storage, covered in Chapters 3 through 5, is finite by definition. A rainwater harvesting system is fundamentally different: correctly designed and maintained, it is a continuously replenished source of water that reduces a household's dependence on both mains supply and static stored reserves. For a household considering any meaningful investment in water preparedness, a rainwater tank connected to roof guttering is consistently one of the highest-return options available, providing everyday utility as well as emergency resilience.

How Much Water a Roof Can Actually Collect

The volume a roof can realistically harvest is a function of three variables: roof catchment area, local rainfall, and a runoff coefficient that accounts for evaporation and the water lost in first-flush diversion. As a working formula:

Annual yield (litres) = roof area (m^2) x annual rainfall (mm) x runoff coefficient

A runoff coefficient of approximately 0.85 is typical for a metal roof, and approximately 0.75 for a tiled roof, reflecting the somewhat greater losses from a tiled surface. As a worked example, a 200 square metre metal roof in a location receiving 1,200 millimetres of rain annually, a figure broadly typical of Sydney, yields approximately 204,000 litres gross before losses, or in the order of 170,000 litres net after accounting for the runoff coefficient.

This figure varies enormously across Australia, and a household in a lower-rainfall inland region should calculate its own figure using local Bureau of Meteorology rainfall data rather than relying on figures calculated for a coastal capital city.

Households should size for the longest realistic dry spell in their specific region, not for the annual average, since a tank sized only for average conditions will run dry precisely in the drought years when it is needed most.

Sizing the Tank to the Purpose

Tank sizing depends heavily on whether rainwater is intended to supplement mains supply or to serve as the household's primary or sole water source.

Urban supplement. For a household connected to mains water and using tank water primarily for the garden, laundry and toilet flushing, with mains water as backup, a tank in the 2,000 to 5,000-litre range is a reasonable and common size. A common rule of thumb used by Australian plumbers is approximately 1,000 litres of tank capacity per household member, with a minimum of 5,000 litres for any household intending to use tank water across laundry, toilets and irrigation.

Meaningful emergency reserve. A household wanting its rainwater tank to function as a genuine emergency reserve, not merely a garden supplement, should aim toward the upper end of this range or beyond, in the order of 5,000 to 10,000 litres, sized against the household's calculated water requirement from Chapter 2 for the number of days of self-sufficiency desired.

Rural or off-grid primary supply. For a property where rainwater is the entire water supply rather than a supplement to mains, sizing needs to be substantially larger again. Government guidance for rural properties relying solely on rainwater suggests planning around 22,500 litres of tank capacity per person per year at typical rural consumption rates, which for a household of four translates to total storage in the order of 90,000 litres or more, sized against the longest historical dry period the region has recorded rather than an average year.

The First Flush Diverter: The Single Most Important Component

A first flush diverter is a device fitted at the tank inlet that automatically diverts the initial flow of rainwater, which carries the highest concentration of accumulated dust, bird and animal droppings, and other debris from the roof, away from the tank before allowing cleaner subsequent flow through to storage.

Sizing. The standard rule of thumb is approximately one litre of diverter capacity per 25 square metres of roof area, meaning the initial 20 to 25 litres of any rainfall event on a typical roof is diverted away rather than collected.

Regulatory status. Queensland's Development Code mandates a first flush diverter wherever rainwater is used for indoor purposes, and equivalent requirements or strong recommendations apply in most other Australian jurisdictions. Even where not strictly mandated, a first flush diverter should be considered a non-negotiable component of any system intended to supply drinking water.

Maintenance. The diverter itself requires periodic cleaning, since the debris it is designed to capture will otherwise accumulate within the device and reduce its effectiveness over time.

Screens, Guards and Mosquito Proofing

Beyond the first flush diverter, several further physical barriers protect water quality and public health.

Gutter mesh. A linear screen fitted along the gutter prevents leaves, sticks and larger debris from entering the downpipe and, eventually, the tank.

Leaf-shedding rain heads. Where gutter mesh is not fitted, a rain head at the top of the downpipe uses a sloping screen to shed leaves before water enters the pipe.

Tank inlet screens. A fine mesh screen at the tank inlet itself catches any remaining debris and prevents small animals, insects and birds from entering the tank directly.

Mosquito proofing. Queensland's Public Health Regulation 2018 requires all rainwater tank openings to be fitted with mesh no coarser than 1 millimetre, or with automatically closing flap valves, specifically to prevent mosquito breeding within the tank. Equivalent requirements exist in other states, and this should be treated as a strict, non-negotiable public health requirement rather than an optional refinement.

Roof Material and Contamination Risk

Not every roof is equally suitable as a drinking water catchment surface, and the following risks should be assessed before committing to a rainwater system intended for potable use.

Lead flashing and old paint. Roofs with lead flashing, particularly common around chimneys and ridges on pre-1970s construction, and roofs painted with pre-1980s lead-based paint, pose a genuine risk of lead contamination in collected water. NSW Health and SA Health both recommend annual testing for lead in tank water from properties with pre-1970s roofing.

Bitumen-based products and treated timber. Rainwater should not be collected from roofs coated with bitumen-based products or from surfaces incorporating exposed preservative-treated timber, both of which can leach chemical contaminants into collected water.

Asbestos roofing. Older asbestos roof sheeting can generally still be used as a rainwater catchment surface provided it has not been coated with lead-based paint or bitumen, but any roof replacement or maintenance work involving asbestos materials should be undertaken by a licensed and appropriately accredited contractor.

Industrial and agricultural proximity. Properties near mining operations, heavy industry, or land subject to regular agricultural spraying should treat rainwater collection with additional caution, since airborne deposits from these activities can settle on the roof catchment area.

Maintenance Schedule

A rainwater system is frequently described in Australian health guidance as needing to be "low maintenance, not no maintenance," and a realistic ongoing maintenance schedule is what separates a genuinely safe drinking water source from a source that merely appears clean.

Weekly to fortnightly, in leaf-drop seasons: Check gutters and the roof catchment area for accumulated leaves and debris.

Monthly to quarterly: Clean the first flush diverter and inlet screens.

Every six to twelve months: Trim overhanging branches, inspect the tank exterior for cracking, staining or algal growth around joints, and confirm mosquito screens remain intact.

Every two to three years: Desludge the tank to remove sediment that accumulates at the base even in a well-maintained system, and inspect the sludge valve, which Australian plumbing standards require on any tank above 500 litres capacity.

Specific circumstances should trigger an immediate inspection outside this routine schedule, including a significant local bushfire that has deposited ash on the roof catchment, aerial spraying of agricultural chemicals nearby, the discovery of a dead animal within the catchment or tank system, or any noticeable change in the water's odour, taste or clarity. Use of the tank for drinking should cease immediately in any of these circumstances until the system has been inspected, flushed and, if necessary, disinfected.

Treatment for Drinking Water Use

Well-maintained rainwater from a clean catchment is, in the assessment of state health authorities including SA Health and NSW Health, generally safe to drink without additional treatment for most healthy users. This assessment comes with an important qualification that households should take seriously rather than treat as a formality: disinfection, through boiling or the installation of a continuous ultraviolet treatment unit, should be considered for anyone in the household who is immunocompromised, very young, elderly, pregnant, or undergoing treatment such as chemotherapy. Where any household member falls into one of these categories, treating all tank water intended for drinking as a matter of routine, rather than only during periods of visible concern, is the more cautious and appropriate approach.

For households seeking mains-equivalent water quality from a rainwater system used for all household purposes, a staged treatment train of sediment filtration, carbon filtration and ultraviolet disinfection, sized to comply with the relevant Australian Standard for treated water quality, provides the highest level of assurance and is increasingly common in new rainwater installations across Australia.

Regulatory Variation by State

Rainwater harvesting regulation in Australia is set at state and territory level, with meaningful variation, and a household should always confirm current requirements with their local council before installation, since council-level rules can add to or override the general state position.

Queensland mandates first flush diverters for indoor use under its Development Code and mosquito-proofing under the Public Health Regulation 2018.

New South Wales and **South Australia** both provide detailed health department guidance on rainwater use and testing, with South Australia generally regarded as having a more permissive regulatory position on potable use than some other states.

Victoria requires new class 1 residential developments to include a minimum 2,000-litre tank connected to at least 50 square metres of roof catchment and plumbed to toilets, as part of its broader water efficiency requirements for new housing.

Tasmania requires a minimum 40-millimetre sludge valve on any tank above 500 litres under AS/NZS 3500.1 and provides specific chlorine dosing guidance through its Department of Health.

Western Australia is generally regarded as having the most restrictive regulatory position on rainwater for potable use among the states.

Getting Started

For a household considering a first rainwater tank, or expanding an existing system toward genuine emergency-reserve capacity, the following sequence is a reasonable approach.

- Calculate realistic annual yield using the roof area, local rainfall and runoff coefficient formula set out earlier in this chapter.
- Decide whether the system is intended as an urban supplement, a meaningful emergency reserve, or a primary rural supply, and size the tank accordingly.
- Confirm current state and local council requirements before purchase or installation.
- Specify a first flush diverter, inlet screen and mosquito-proof mesh as non-negotiable inclusions, regardless of whether local regulation strictly mandates them.
- Engage a licensed plumber for any connection to household plumbing, as required in every Australian jurisdiction.
- Establish the maintenance schedule set out in this chapter from the day the system is commissioned, rather than treating it as a future task.

Sources referenced in this chapter

Australian Government, Rainwater, yourhome.gov.au/water/rainwater

SA Health, Rainwater, sahealth.sa.gov.au

Queensland Public Health Regulation 2018, mosquito-proofing and rainwater tank requirements

Tasmanian Department of Health, drinking water quality and chlorine dosing guidance, via Consumer, Building and Occupational Services (Tasmania)

Department of Health Western Australia, A Compilation of Australian Standards on Water Holding Tanks

enHealth, Guidance on the Use of Rainwater Tanks

CHAPTER 7

Wells, Bores and Digging for Water

Groundwater, the water stored beneath the surface in soil and rock formations known as aquifers, is the second major renewable water resource available to many Australian properties, alongside the rainwater harvesting covered in Chapter 6. This chapter covers two very different approaches to reaching it: the properly licensed bore, planned and constructed as permanent household infrastructure, and the primitive emergency well, dug by hand as an urgent, short-term response when no other water source is available. The two should not be confused with one another. A primitive emergency well is a survival technique for a genuine emergency. A licensed bore is a substantial piece of household infrastructure requiring professional construction and ongoing regulatory compliance.

Understanding Groundwater and Aquifers

An aquifer is a layer of permeable rock, sand or gravel that holds and transmits groundwater. Aquifers are recharged over time by rainfall percolating down through the soil, and the depth at which water is found, known as the water table, varies enormously across Australia depending on local geology, rainfall and the extraction history of the surrounding area.

Groundwater quality varies just as much as its depth and yield. Some Australian aquifers hold water suitable for drinking with minimal treatment. Others hold water too saline for household use, or water carrying naturally occurring minerals, agricultural runoff, or contamination from nearby land use at concentrations that make treatment essential before consumption. This variation is precisely why every state requires groundwater bores to be constructed by a licensed driller and, in most cases, requires the resulting water to be tested before it is relied upon.

Licensed Bores: Planning as Permanent Infrastructure

For a household or rural property considering a bore as a planned, permanent water source rather than an emergency measure, the process runs through several stages that are broadly consistent across Australian states, though the specific licensing bodies and terminology vary.

Approval before drilling. Every Australian state and territory requires a formal approval, permit or licence before any bore construction begins. In New South Wales this is a water supply work approval issued by WaterNSW; in South Australia it is a Well Permit issued by the Department for Environment and Water; in Victoria it is a Bore Construction Licence under the Water Act 1989; in Queensland it involves confirming development approval requirements with the local council and relevant state authority. In every case, drilling before the appropriate approval is in hand exposes both the landholder and the driller to legal risk.

Engaging a licensed driller. All states require bore construction to be carried out by a driller holding the appropriate licence class for the aquifer type involved. Licence classes typically distinguish between single-aquifer bores, multiple-aquifer bores, and the more complex requirements of artesian conditions, reflecting the genuinely different skills and risks involved in each. Engaging an unlicensed operator, even where a friend or contractor offers to do the work informally, creates a real risk of aquifer contamination, cross-contamination between aquifers, and a bore that fails to perform as expected.

Site assessment. Before drilling, an assessment of expected depth, water quality and yield at the specific site helps establish whether the investment is likely to be worthwhile. Yield and depth can vary significantly even between adjoining properties, and a driller with local experience is generally better placed to advise on this than general published guidance.

Construction and disinfection. All bores must be constructed to the Minimum Construction Requirements for Water Bores in Australia, a national technical standard maintained by the Australian Drilling Industry Association, covering casing, sealing between aquifers, and headworks design. On completion, new bores are typically disinfected with a sterilising agent to reduce the risk of iron bacteria establishing itself within the bore, since these bacteria are extremely difficult to eradicate once present.

Testing and reporting. After construction, the driller tests the bore's yield and, ideally, its water quality, and submits a completion report to the relevant state authority, generally within 28 to 60 days depending on jurisdiction. A copy of this report, along with the drill log, should be kept by the property owner as a permanent record.

Extraction licensing. Constructing a bore and extracting water from it are frequently governed by separate approvals. In Victoria, for example, the initial Works Licence covers construction, while ongoing extraction for domestic and stock purposes is generally then automatically permitted, but any commercial use, including significant irrigation, requires a further Licence to Take and Use Water, which is capped or restricted in some areas.

Pumps and Power for Bore Water

A bore is only as useful as the system that brings water to the surface, and this system introduces the single most common failure point for bore-supplied households during a wider disruption: dependence on grid electricity.

Submersible electric pumps. The most common pump type for domestic bores, submersible pumps are efficient and reliable under normal conditions but fail immediately during a power outage unless supported by an alternative power source.

Solar bore pumps. An increasingly common option for rural and off-grid properties, solar-powered pumps remove the dependency on grid power and mains-connected inverters, though they still depend on functioning solar panels and controllers, and typically pump at a reduced rate during periods of poor sunlight.

Backup power planning. A household relying on a bore as a primary water source should treat backup power for the pump, whether through a generator, a battery and inverter system, or a dedicated solar pump, as an essential companion to the bore itself rather than an optional extra. A bore with no means of extracting water during a blackout provides no practical benefit during exactly the kind of disruption this book is concerned with.

Ongoing Water Quality Management

Bore water quality is not static, and should be reassessed periodically rather than assumed safe indefinitely on the strength of the initial test.

Test for bacterial contamination, including coliform bacteria, on a schedule recommended by the relevant state health authority, and immediately after any flood event that may have introduced surface contamination into the bore.

Test for salinity and relevant chemical parameters, particularly in agricultural areas or regions with a known history of naturally elevated mineral content.

Monitor for a decline in yield over time, which can indicate either a falling water table in the surrounding aquifer or a mechanical problem within the bore itself, and have any significant, sustained decline investigated by a licensed driller.

Primitive Emergency Wells

Where a household or an individual has no access to a licensed bore, no rainwater tank, and no stored reserve, and is facing a genuine emergency in a location where groundwater is likely to be present near the surface, a hand-dug emergency well is a legitimate short-term survival technique. It is not a substitute for the planned infrastructure described above, and water obtained this way should always be treated before drinking, using the methods set out in Part VI.

Reading the landscape for a viable dig site. The chance of success depends heavily on site selection.

Look for a dry creek bed or the lowest point of a dry watercourse, where sub-surface moisture is most likely to persist even after surface water has disappeared.

Look for a concentration of green, healthy vegetation in an otherwise dry landscape, which frequently indicates a shallower water table beneath that specific point.

On a coastal or lake shoreline, particularly on a sandy beach or dune, look for a point several metres back from the current waterline, above the high tide mark.

The dry creek bed seep well. In a dry or nearly dry watercourse, digging a hole approximately half a metre to a metre deep at the lowest point of the bed can intercept sub-surface moisture that continues to flow beneath a surface that appears completely dry. Water frequently seeps slowly into a hole of this kind over twenty minutes to several hours, rather than appearing immediately, and patience is part of the technique.

The beach or dune well. On many sandy coastlines, a lens of fresh or brackish water can sit above the denser saltwater intrusion beneath a beach or dune, recharged by rainfall percolating down through the sand. Digging a hole several metres back from the high tide line, generally behind the first line of dunes rather than on the open beach itself, can intercept this fresh water lens. Water drawn this way is frequently brackish rather than fully fresh, should always be treated before drinking, and should never be assumed safe purely because it does not taste strongly of salt.

Safety considerations for hand digging. Any hand-dug hole beyond a shallow depth carries a genuine risk of collapse, particularly in loose sand or unstable soil. Digging should be undertaken with sloped rather than vertical sides wherever the soil allows it, no one should enter a dug hole deeper than their own height without support, and digging should stop and the hole should be abandoned if the walls show any sign of active collapse or slumping.

Treating water from an emergency well. Water obtained through hand digging has had minimal to no natural filtration and may carry biological contamination, dissolved salts, or both. It should always be assessed and treated using the screening methods in Chapter 14 and the treatment methods in Part VI before consumption, and should never be assumed safe on the basis of appearance or taste alone.

Choosing Between the Two Approaches

A licensed bore is an appropriate investment for a rural or semi-rural property planning genuine long-term water self-sufficiency, where the cost of professional drilling, typically running to many thousands of dollars depending on depth and geology, is justified by decades of reliable supply. A primitive emergency well is appropriate only as a genuine emergency response when no other water source is available and professional infrastructure is not an option, and the material in Part VIII of this book returns to this technique in more detail for the specific extreme scenarios in which it becomes relevant.

GROUNDWATER ACCESS: LICENSED BORE vs EMERGENCY WELL

LICENSED BORE
(Permanent Infrastructure)

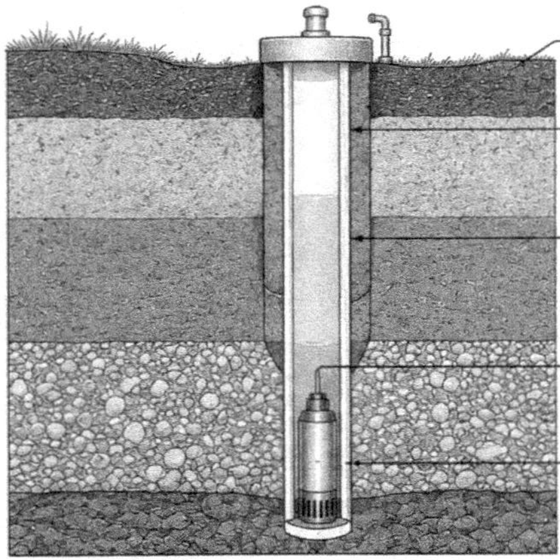

Constructed by licensed driller.
Requires state approval + testing.

EMERGENCY SEEP WELL
(Hand-Dug, Short-Term)

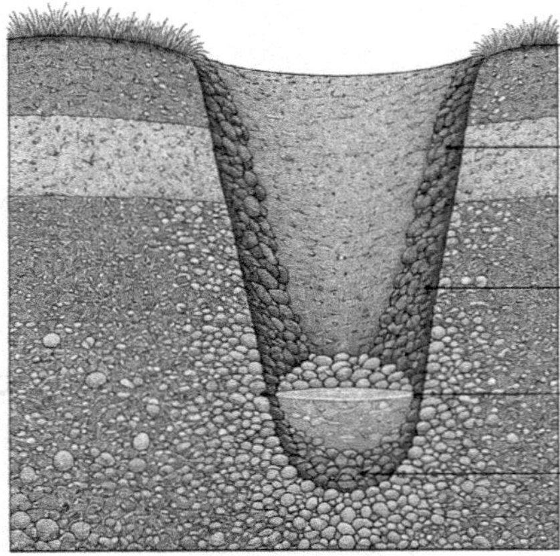

Emergency technique only.
Always screen + treat before drinking.

	Licensed Bore	Emergency Seep Well
PURPOSE	Planned, long-term water supply	Short-term survival in an emergency -
APPROVAL	Requires state approval / licence	No approval (emergency use only)
CONSTRUCTION	By licensed driller to national standards	Hand-dug with basic tools
YIELD	Reliable, higher volume	Low volume, variable
PROTECTION	Cased, sealed, protected from contamination	No casing – vulnerable to contamination
DURATION	Permanent infrastructure	Temporary – use while other options unavailable
WATER QUALITY	Tested – may require treatment	Must be screened and treated before drinking

---

Sources referenced in this chapter

WaterNSW, Drillers Portal, waternsw.com.au

Department for Environment and Water (South Australia), Drilling Wells, environment.sa.gov.au

Business Queensland, Steps to Construct a Bore, business.qld.gov.au

Resources Victoria, Groundwater Licensing and Trading, resources.vic.gov.au

Australian Drilling Industry Association, Minimum Construction Requirements for Water Bores in Australia

Northern Territory Government, water drilling licence classes, nt.gov.au

CHAPTER 8

Reading the Landscape for Water

When stored supplies are exhausted, a rainwater tank has run dry, and no bore is available, the ability to locate water in the surrounding landscape becomes the deciding factor in a genuine survival situation. This chapter sets out the practical indicators used to identify likely water sources across Australian terrain, before Chapter 9 covers atmospheric water and Chapter 10 covers extracting water directly from plants.

An important caution applies throughout this chapter and the two that follow it. Locating water and confirming that it is safe to drink are separate problems. No natural water source in Australia should be assumed safe without the screening and treatment set out in Parts V and VI, regardless of how clear, remote or aesthetically clean it appears.

Terrain and Geological Indicators

Landscape features frequently reveal the presence of groundwater at or near the surface, without any need to dig or drill.

Valley floors and low points. Water moves downhill through soil in the same way it does across the surface. The lowest point of any valley or depression is more likely to have a shallow water table than the surrounding higher ground.

Rock outcrops and cliff bases. Water percolating through porous rock frequently emerges at the base of a rock face or outcrop, particularly where a harder, less permeable rock layer sits beneath a more porous one, forcing water sideways to the surface at the boundary between the two layers.

Dry creek beds. Even a watercourse that appears completely dry on the surface frequently retains moisture beneath the surface at its lowest points, as covered in the seep well technique in Chapter 7.

Changes in vegetation density and colour. A patch of noticeably greener, denser or larger vegetation within an otherwise dry landscape is one of the most reliable visual indicators of a shallower water table beneath that specific location.

Sand and gravel deposits at the base of slopes. These deposits frequently indicate a natural drainage path where water collects and moves through the subsurface even when no surface flow is visible.

Indicator Plants in the Australian Landscape

Certain plant species and vegetation patterns are strongly associated with the presence of water, either because the species itself requires a permanently available water source or because its presence indicates favourable soil moisture conditions more broadly.

River red gums (Eucalyptus camaldulensis). Found across much of inland Australia, this species grows along watercourses and floodplains and rarely survives far from a reliable water source, whether that source is currently flowing on the surface or present as groundwater beneath a dry channel.

Paperbarks (Melaleuca species). Commonly found in swampy, low-lying or seasonally inundated ground across much of eastern and northern Australia, paperbarks are a strong indicator of consistently moist or waterlogged soil.

Reeds, sedges and rushes. Dense stands of these plants, particularly in a low-lying area, indicate consistently moist soil at or very near the surface, and frequently surround a spring, soak or semi-permanent waterhole even when no open water is immediately visible.

A sudden band of green in an otherwise dry landscape. This general pattern, rather than any single species, is often the most useful indicator available in unfamiliar country. A visible line or patch of markedly greener vegetation crossing an otherwise dry hillside frequently follows a subsurface water path.

Animal Behaviour as an Indicator

Native and introduced animals in Australia depend on water just as consistently as plants do, and their behaviour patterns provide a second, independent layer of evidence.

Converging animal trails. In dry terrain, well-used trails made by kangaroos, wallabies, cattle or other animals frequently converge toward a single point, and following several converging trails downhill, particularly in the direction they appear most heavily used, will often lead to a water source.

Bird activity in the early morning and late afternoon. Many bird species, particularly flocking species such as parrots and pigeons, travel to water sources at consistent times of day, typically shortly after dawn and again before dusk. Observing the direction of consistent bird flight at these times, over more than one day if possible, can indicate the general direction of a water source even before it is visible.

Insect activity. A concentration of bees, in particular, is a reasonably strong indicator of water within a relatively short flying distance, since bees require a regular water source for hive cooling as well as drinking.

Fresh animal tracks and dung near a specific point, particularly where multiple trails intersect, frequently indicate a soak, spring or waterhole close by, even where no open water is immediately visible from that point.

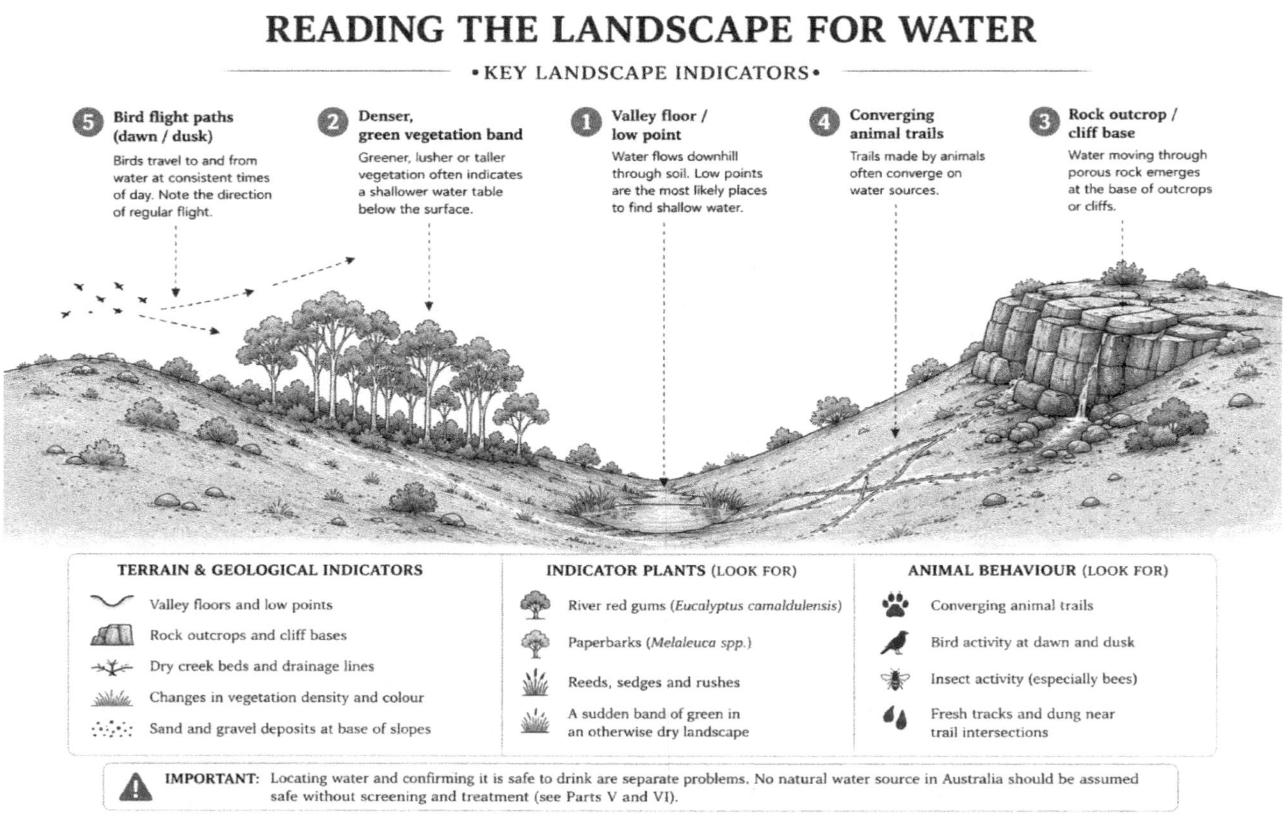
READING THE LANDSCAPE FOR WATER
• KEY LANDSCAPE INDICATORS •
5 Bird flight paths (dawn / dusk)
Birds travel to and from water at consistent times of day. Note the direction of regular flight.
2 Denser, green vegetation band
Greener, lusher or taller vegetation often indicates a shallower water table below the surface.
1 Valley floor / low point
Water flows downhill through soil. Low points are the most likely places to find shallow water.
4 Converging animal trails
Trails made by animals often converge on water sources.
3 Rock outcrop / cliff base
Water moving through porous rock emerges at the base of outcrops or cliffs.
TERRAIN & GEOLOGICAL INDICATORS
Valley floors and low points
Rock outcrops and cliff bases
Dry creek beds and drainage lines
Changes in vegetation density and colour
Sand and gravel deposits at base of slopes
INDICATOR PLANTS (LOOK FOR)
River red gums (Eucalyptus camaldulensis)
Paperbarks (Melaleuca spp.)
Reeds, sedges and rushes
A sudden band of green in an otherwise dry landscape
ANIMAL BEHAVIOUR (LOOK FOR)
Converging animal trails
Bird activity at dawn and dusk
Insect activity (especially bees)
Fresh tracks and dung near trail intersections
IMPORTANT: Locating water and confirming it is safe to drink are separate problems. No natural water source in Australia should be assumed safe without screening and treatment (see Parts V and VI).

Dowsing: What the Evidence Actually Shows

Water divining, also known as dowsing or water witching, is a long-standing practice in which a person walks across land holding a forked stick, a pendulum, or a pair of L-shaped rods, which are said to move or cross when the dowser passes over underground water. It remains widely practised in parts of rural Australia and elsewhere, and this book addresses it directly rather than omitting it, because an honest account of what the evidence actually shows is more useful than either an uncritical endorsement or a dismissal that ignores why the practice persists.

Every controlled scientific study of dowsing, including formal double-blind trials conducted in Germany, the United States and elsewhere, has found that dowsers locate water no more reliably than random chance under conditions where the actual location of the water is unknown to the dowser and not detectable through ordinary sensory cues (United States Geological Survey). The movement of the rod or stick itself is well explained by the ideomotor effect, in which small, unconscious muscle movements by the person holding the rod are amplified by the unstable way the tool is held, producing motion that feels externally caused but is generated entirely by the dowser.

The practice appears to succeed in many real-world cases for two separate reasons that have nothing to do with any special ability. First, in regions with adequate rainfall and favourable geology, groundwater is present at a drillable depth across almost the entire landscape, so a bore sunk at a location chosen by dowsing, or chosen at random, or chosen by no method at all, will frequently strike water regardless of the method used to select the spot. Second, an experienced local dowser often has genuine, if unconscious, knowledge of the same terrain and vegetation indicators set out earlier in this chapter, and may be responding to that real information while attributing the result to the rod itself.

The practical conclusion for a preparedness context is straightforward: dowsing should not be relied upon as a primary method for locating water in a genuine survival situation, and the terrain, vegetation and animal indicators set out in this chapter, which are based on observable, physical evidence, should be used instead. Readers who wish to use dowsing alongside these methods, whether from personal conviction or family tradition, are not likely to be harmed by doing so provided it supplements rather than replaces the evidence-based indicators in this chapter.

Assessing a Located Source Before Approaching It

Once a probable water source has been identified through any of the methods above, a brief visual assessment before collection helps avoid the least safe sources when an alternative is available nearby.

- Check for any dead animals in or immediately around the water, which indicates a meaningful risk of biological or chemical contamination.

- Check for algal bloom, appearing as a greenish, bluish, brown or reddish scum or paint-like coating on the surface, often with a musty smell. Water showing any sign of algal bloom should be avoided entirely, since the toxins produced by blue-green algae are not removed by boiling or standard filtration.

- Where a choice exists, prefer flowing water over still water, and prefer water collected upstream of any visible livestock access, road crossing or human habitation.

Note the general condition of the surrounding vegetation and soil, since a source surrounded by obviously unhealthy or stressed vegetation, or by heavily eroded, bare ground from livestock trampling, is more likely to carry a heavier contamination load than a source in otherwise healthy country.

None of these observations replace the formal screening and treatment set out in Parts V and VI. They simply help direct effort toward the most promising and least risky source available when more than one option exists.

Sources referenced in this chapter

United States Geological Survey (USGS), Water Dowsing, usgs.gov

Skeptical Inquirer, Testing Dowsing: The Failure of the Munich Experiments

BBC Science Focus, Is There Any Scientific Evidence for Dowsing?

General Australian native flora references on river red gum and paperbark habitat association

DOWSING: WHAT THE EVIDENCE ACTUALLY SHOWS

1. WHAT IS DOWSING?

A person walks across land holding a tool such as a forked stick, pendulum or L-shaped rods. The tool is said to move or cross when passing over underground water.

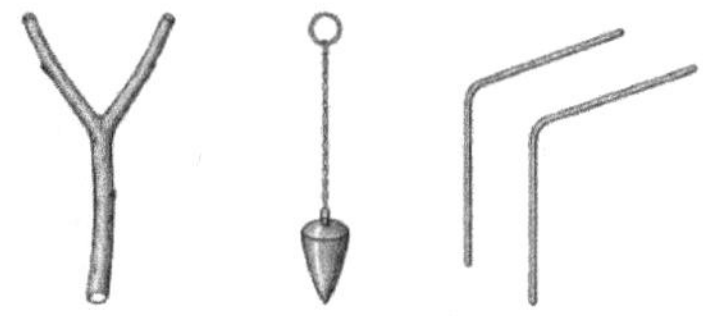

2. WHAT THE SCIENCE SHOWS

Every controlled scientific study of dowsing, including double-blind trials in Germany, the United States and elsewhere, has found that dowsers locate water no more reliably than random chance when the location of water is unknown and not detectable through ordinary senses.

The movement of the rod or stick is explained by the *ideomotor effect*: small, unconscious muscle movements by the dowser are amplified by the unstable way the tool is held, creating motion that feels externally caused but is generated by the dowser.

3. WHY DOWSING APPEARS TO WORK

REASON 1: WATER IS WIDESPREAD

In many regions with adequate rainfall and favourable geology, groundwater exists at a drillable depth across almost the entire landscape. A bore drilled by dowsing, at random, or by no method at all will often strike water.

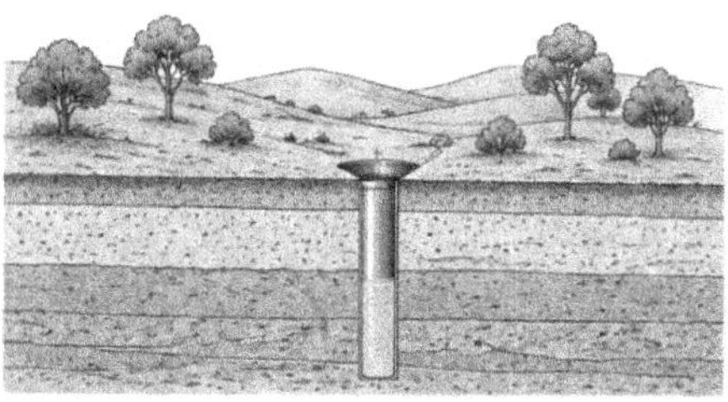

REASON 2: UNCONSCIOUS KNOWLEDGE

Experienced local dowsers often have genuine, unconscious knowledge of the same terrain, vegetation and animal indicators. They may be responding to that real information while attributing the result to the rod.

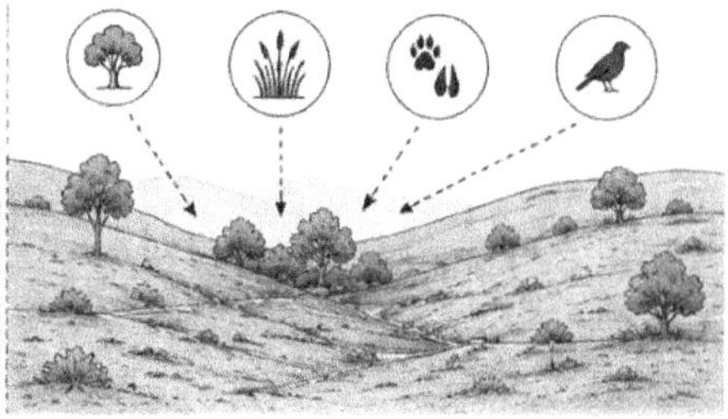

4. PRACTICAL CONCLUSION

- ✗ Do not rely on dowsing as a primary method for locating water in a survival situation.
- Use terrain, vegetation and animal indicators based on observable, physical evidence.
- ✓ Dowsing may be used alongside these methods if personally preferred or traditional, provided it supplements rather than replaces evidence-based indicators.

EVIDENCE-BASED INDICATORS

- Valley floors / low points
- Rock outcrops / cliff bases
- Dry creek beds / drainage lines
- Greener, denser vegetation
- Sand / gravel deposits
- Converging animal trails
- Bird flight paths (dawn / dusk)
- Insect activity (e.g. bees)

IMPORTANT: Locating water and confirming it is safe to drink are separate problems. No natural water source in Australia should be assumed safe without screening and treatment (see Parts V and VI).

CHAPTER 9

Dew, Condensation and Solar Stills

Water suspended in the atmosphere, whether as dew, fog or humidity, and water held within damp soil, represent a further source available even in landscapes where no visible surface water exists. The methods in this chapter typically produce modest volumes compared to a rainwater tank or a located spring, and should be understood as a supplementary technique for extending an existing supply, or as one part of a combined strategy in a genuine survival situation, rather than a primary source capable of meeting a household's full daily requirement on its own.

Dew Collection

Dew forms overnight as moist air cools and condenses on cooler surfaces, particularly vegetation, metal and rock, and is most reliable in the hours immediately before and after dawn, before rising temperatures cause it to evaporate.

- Identify a location with dense, low vegetation, since grass and low shrubs typically hold more dew per unit area than bare ground or taller, sparser vegetation.
- Use an absorbent cloth, a clean t-shirt or a bandana, and drag it across dew-covered grass or leaves in the early morning, before the sun has risen high enough to begin evaporating the moisture.
- Wring the cloth into a clean container. Repeat across as wide an area as practical.
- Under good conditions, this technique can collect several

hundred millilitres per hour, though the actual yield depends heavily on humidity, vegetation density and the size of the area covered.

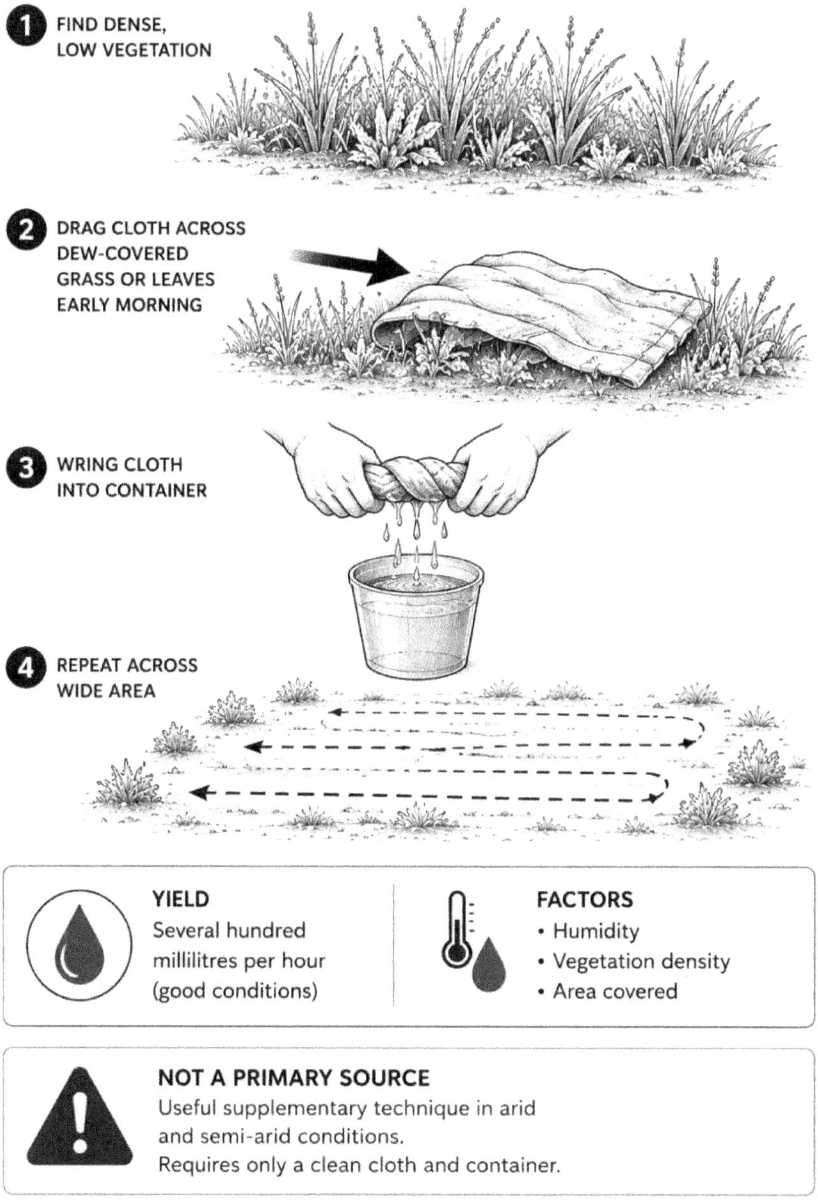

Dew collection should never be treated as a primary water source. It is a genuinely useful supplementary technique in arid and semi-arid Australian conditions, where morning dew can be more reliable than any other atmospheric source, and it requires no equipment beyond a clean cloth and a container.

Fog and Condensation Harvesting

In coastal, mountain or early-morning foggy conditions, moisture suspended in fog can be captured using a simple mesh or fabric surface that intercepts fine water droplets as fog moves through it.

A length of fine mesh, shade cloth or closely woven fabric, stretched vertically between two supports and angled slightly, allows fog to condense on its surface and run down into a collection trough or container placed beneath it.

This technique is most effective in locations with reliably foggy conditions, such as elevated coastal country or mountain ranges affected by regular cloud cover, and is of limited use in the dry interior of the continent, where fog is rare.

Vehicle windscreens, tarpaulins and any other broad, smooth surface exposed to fog or heavy overnight condensation can be wiped down in the same way as dew-covered vegetation, using a clean cloth, to recover a further modest quantity of water.

1. STRETCH FINE MESH OR FABRIC BETWEEN TWO SUPPORTS, ANGLED SLIGHTLY.

2. FOG CONDENSES ON THE SURFACE AND RUNS DOWN INTO TROUGH / CONTAINER.

3. MOST EFFECTIVE IN FOGGY, COASTAL OR MOUNTAIN LOCATIONS. LIMITED USE IN DRY INTERIOR.

4. WIPE DOWN VEHICLE WINDSCREENS, TARPS OR SMOOTH SURFACES WITH A CLEAN CLOTH TO COLLECT MOISTURE.

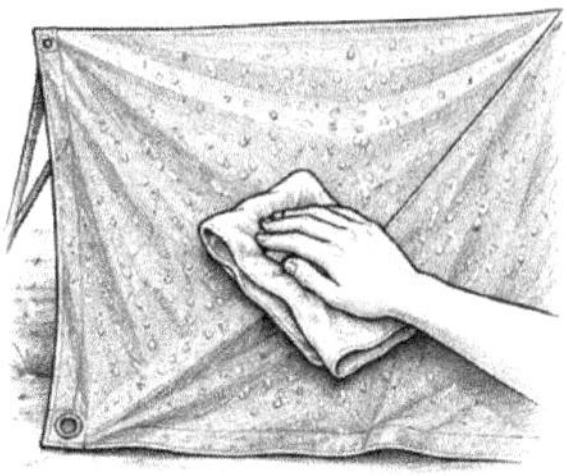

CAPTURES MOISTURE FROM FOG	SIMPLE SETUP: MESH + SUPPORTS + CONTAINER	BEST IN FOGGY, COASTAL OR MOUNTAIN AREAS	USE SMOOTH SURFACES TO RECOVER MORE WATER

The Transpiration Bag

A transpiration bag exploits the natural process by which living plants release water vapour through their leaves, and is a genuinely useful technique in vegetated country, including much of temperate and subtropical Australia.

- Select a leafy, healthy branch or a low shrub in full sun, avoiding any plant known or suspected to be toxic, since the water produced can carry trace compounds from the plant itself.

- Enclose the branch or shrub in a clear plastic bag, leaving as much leaf surface as possible inside the bag without crushing it.

- Seal the bag around the branch or at the base of the shrub as tightly as possible, and position the bag so one corner sits at the lowest point, allowing condensed water to pool there.

- Leave the bag in full sun for several hours. Water vapour released by the leaves condenses on the inside of the bag as it cools, running down to the low point.

- Carefully collect the pooled water without disturbing the seal more than necessary, and consider adding fresh foliage if the bag is to be left in place for an extended period, since transpiration slows once the enclosed air reaches saturation and the leaves can no longer release further moisture effectively.

This technique typically produces a modest but genuinely useful volume, particularly when several bags are set up simultaneously on different plants.

1 SELECT PLANT

- Leafy, healthy branch or low shrub
- In full sun
- Avoid toxic plants

2 ENCLOSE IN PLASTIC BAG

- Use clear plastic bag
- Include as much leaf surface as possible

3 SEAL AND POSITION

- Seal tightly around branch/base
- Lowest corner points down

4 LEAVE IN FULL SUN

- Several hours
- Vapour from leaves condenses on inside of bag and runs downward

5 COLLECT WATER

- Carefully collect pooled water
- Add fresh foliage if leaving for extended period

YIELD
Modest but genuinely useful.

BEST PRACTICE
Set up several bags on different plants for better total yield.

Solar Stills

A solar still uses the sun's heat to evaporate moisture from damp soil, plant material or contaminated water, then captures the resulting condensation on the underside of a sheet of clear plastic, producing water that is distilled and therefore free of the salts, heavy metals, bacteria and other non-volatile contaminants present in the original source.

Building a below-ground solar still:

- Dig a hole approximately 60 centimetres wide and 45 centimetres deep, in a sunny location with the dampest soil available. A dry creek bed, as identified in Chapter 8, is an excellent site.

- Place a clean container at the centre of the base of the hole to collect condensed water.

- Where available, add cut vegetation, or contaminated or salt water, around the container within the hole to increase the moisture available for evaporation. Adding urine to the pit is a legitimate technique in a genuine survival situation, since the distillation process leaves the waste products behind and recovers only the water vapour.

- Cover the hole with a sheet of clear plastic, secured tightly around the edges with soil or rocks so no vapour can escape.

- Place a small weight, such as a smooth stone, at the centre of the plastic sheet directly above the container, pulling the plastic into a shallow cone shape.

- As the sun heats the air and soil within the sealed pit, moisture evaporates and condenses on the underside of the cooler plastic sheet, running down the slope of the cone and dripping into the container below.

- A short length of flexible tubing, run from the container to outside the sealed edge of the plastic, allows water to be drawn off without breaking the seal and interrupting the process.

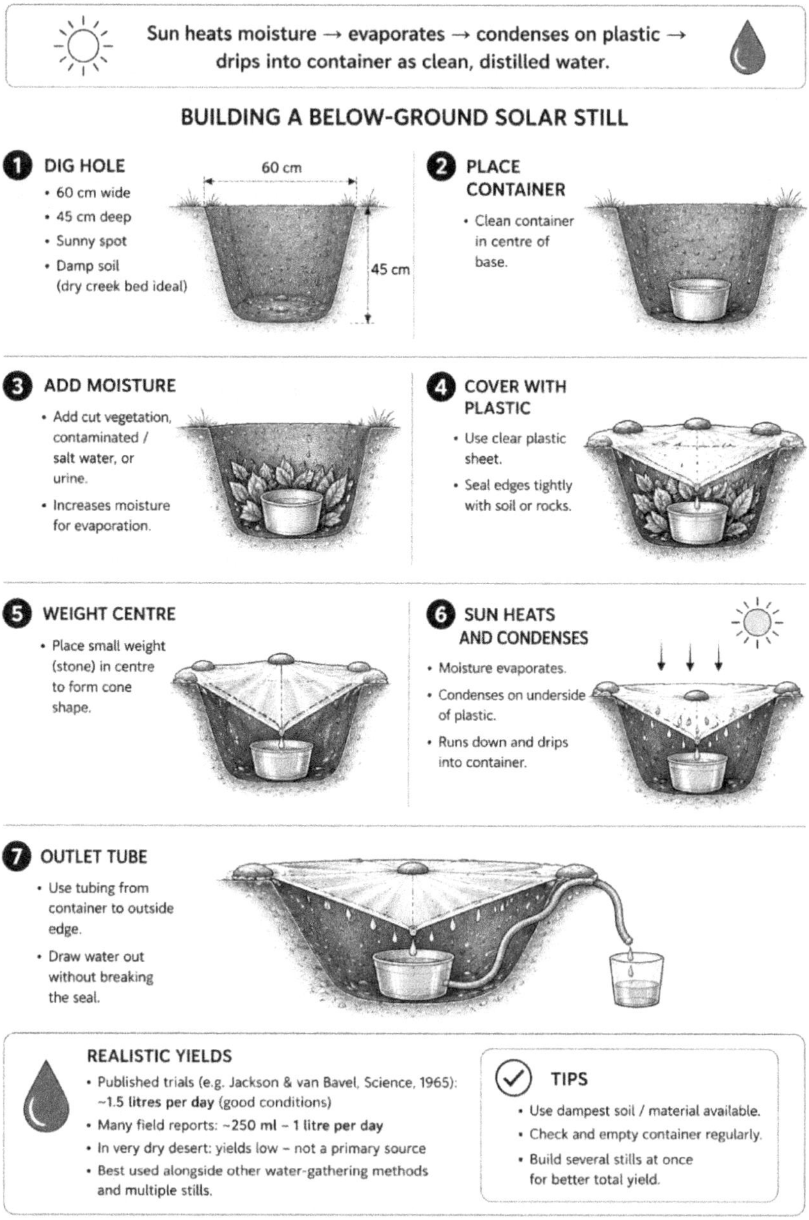

Realistic expectations. Published field trials of this technique, including a widely cited 1965 study in the journal Science, recorded yields in the order of 1.5 litres per day from a single well-constructed still under favourable conditions (Jackson and van Bavel, Science, 1965). Many field reports, however, describe considerably lower yields, often in the range of 250 millilitres to one litre per day, depending heavily on soil moisture, available sunlight and the quality of the seal. In genuinely dry desert conditions, where daily water requirement can exceed 3.8 litres per person, a single still will typically fall well short of meeting that requirement on its own, and the effort and moisture expended in digging the pit itself should be weighed against the modest and uncertain return. A solar still is best used as one component of a broader strategy, ideally alongside the landscape-reading techniques from Chapter 8, rather than as a stand-alone solution, and building several stills simultaneously improves the household or individual's total yield more reliably than relying on a single unit.

The floating still for marine survival. A specific still design, developed for military use and still found in some life raft survival kits, uses an inflatable floating outer bladder with a smaller inner bag suspended inside it. Seawater is poured into the inner bag, and the outer bladder, floating on the surrounding seawater, provides both flotation and the cooling surface needed for condensation. Reported output for these purpose-built units ranges from approximately 1.4 to 2.4 litres per day, considerably better than an improvised still, reflecting the benefit of a purpose-built seal and surface area.

Purpose-built still used in military and life raft kits.

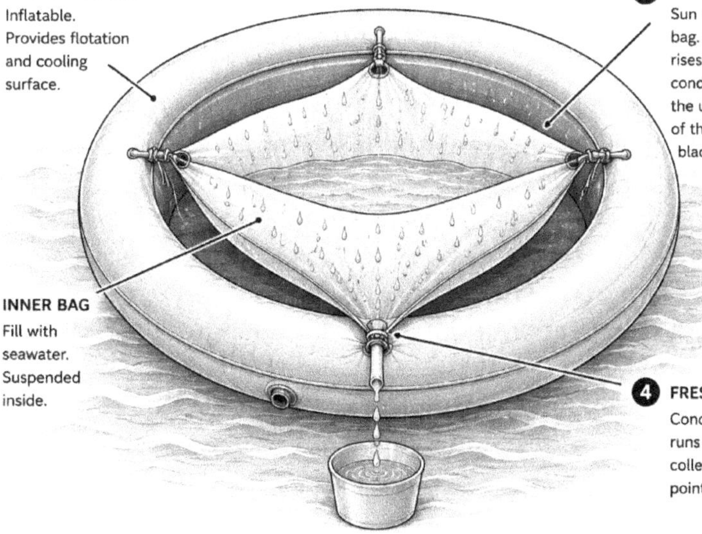

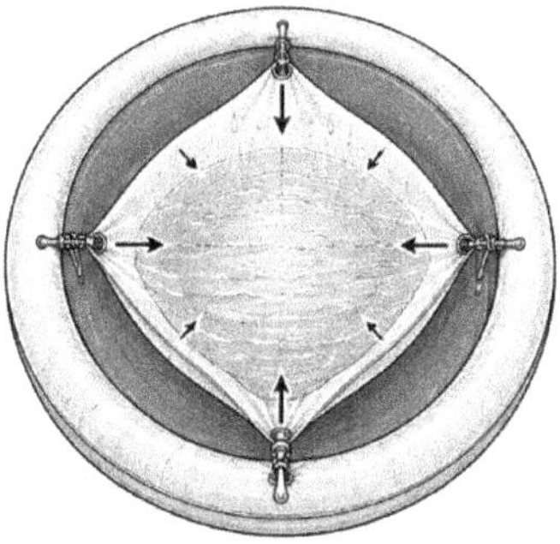

1. **INFLATE OUTER BLADDER**
   Ensure airtight seal.
2. **FILL INNER BAG WITH SEAWATER**
   Tie securely and suspend inside bladder.
3. **PLACE IN SUNLIGHT**
   Sun heats seawater; vapour rises.
4. **COLLECT FRESH WATER**
   Condensation runs down and drips from low point.

OUTPUT

Approximately **1.4 – 2.4 litres per day**

Better than improvised stills.

BEST USED

- Marine survival situations
- Calm seas
- Strong sunlight
- Regularly check and empty collected water

Combining Atmospheric Techniques

None of the methods in this chapter are efficient enough to serve as a household's primary water source under normal circumstances, but in combination, and particularly in a genuine survival situation where every source matters, they meaningfully extend the time available before a more substantial source, whether stored, harvested or located through the methods in Chapters 6 through 8, becomes essential. A person managing dew collection at dawn, a transpiration bag through the day, and a solar still running continuously in the background is drawing on three independent atmospheric and soil moisture sources at once, each requiring minimal equipment and none competing with the others for the same window of time.

Sources referenced in this chapter

Jackson, R.D. and van Bavel, C.H.M., Solar Distillation of Water from Soil and Plant Materials: A Simple Desert Survival Technique, Science, 1965

Safe Drinking Water Foundation, Solar Water Distillation, safewater.org

DesertUSA, Desert Survival: How to Build a Solar Still, desertusa.com

CHAPTER 10

Extracting Water from Plants

Plants store water in roots, stems, fruit and specialised tissue, and in the right country this stored water can be extracted directly, without any dependence on rainfall, soil moisture or a located spring. This chapter sets out the specific techniques and species relevant to Australian conditions, drawing on both traditional Aboriginal knowledge developed over tens of thousands of years of managing exactly this problem, and the more familiar international techniques involving cacti and coconuts that many readers will already associate with plant-based water extraction.

The First Rule: Plant Identification Before Extraction

Before extracting water from any plant, correct identification is essential. Milky, coloured or strongly odorous sap is a general, though not absolute, warning sign of toxicity, and any plant that cannot be confidently identified should not be used as a water source. This caution matters more, not less, in a genuine emergency, since the physiological stress of dehydration is not improved by adding plant toxin exposure or gastrointestinal illness on top of it. Where in doubt, prioritise a technique that does not require ingesting the plant material itself, such as the transpiration bag covered in Chapter 9, over direct extraction from an unfamiliar species.

Water-Bearing Roots: The Central Australian Technique

Aboriginal peoples across the arid and semi-arid interior of Australia developed a highly refined technique for extracting water from the roots of specific tree species, and this remains the most significant plant-based water source documented in Australian conditions.

Mallee eucalypts (including White Mallee, *Eucalyptus dumosa*, and Red Mallee, *Eucalyptus oleosa*) were historically the most important source, particularly across the dry country of central Australia and western New South Wales. These species grow as scattered small trees or dense thickets across much of Australia's arid zone and store water within broad conduits in their root system, particularly during spring and summer when sap flow is highest.

The extraction technique. Beginning roughly a metre out from the trunk, a root is exposed by digging down, typically no deeper than 25 centimetres. Once located, the root is chopped free and pulled from the ground, often 12 to 24 metres of continuous root in a single extraction. The root is then cut into shorter lengths and stood upright in a container, allowing water to drain from the cut ends. A single root of this kind can yield 370 millilitres or more.

Needlewood (*Hakea leucoptera*), **Coolibah** (*Eucalyptus microtheca*), and various **kurrajong and bottle tree species** (*Brachychiton* species, including the Desert Kurrajong, *Brachychiton gregorii*) are further documented water-bearing species across the arid interior, identifiable by their broad sap conduits and, in the case of bottle trees, a distinctively swollen trunk that stores water directly within the tree itself.

Desert Oak (*Allocasuarina decaisneana*) stores water both in root systems and within cavities in the trunk. These trunk cavities, with small entrance holes that exclude larger animals, frequently hold genuinely good-quality water, and Aboriginal trackers historically used the presence of ants around a small opening as an indicator of a water-holding cavity within.

This technique requires practice, physical effort and, ideally, knowledge passed down specifically for the local country, since identifying which individual trees within a species are actually water-bearing at a given time is a skill in its own right. It should be understood as a genuine emergency technique for the extreme scenarios covered in Part VIII, not a routine water source, and readers travelling or living in remote arid country are encouraged to seek local Aboriginal knowledge and guidance directly wherever possible, rather than relying solely on a written description.

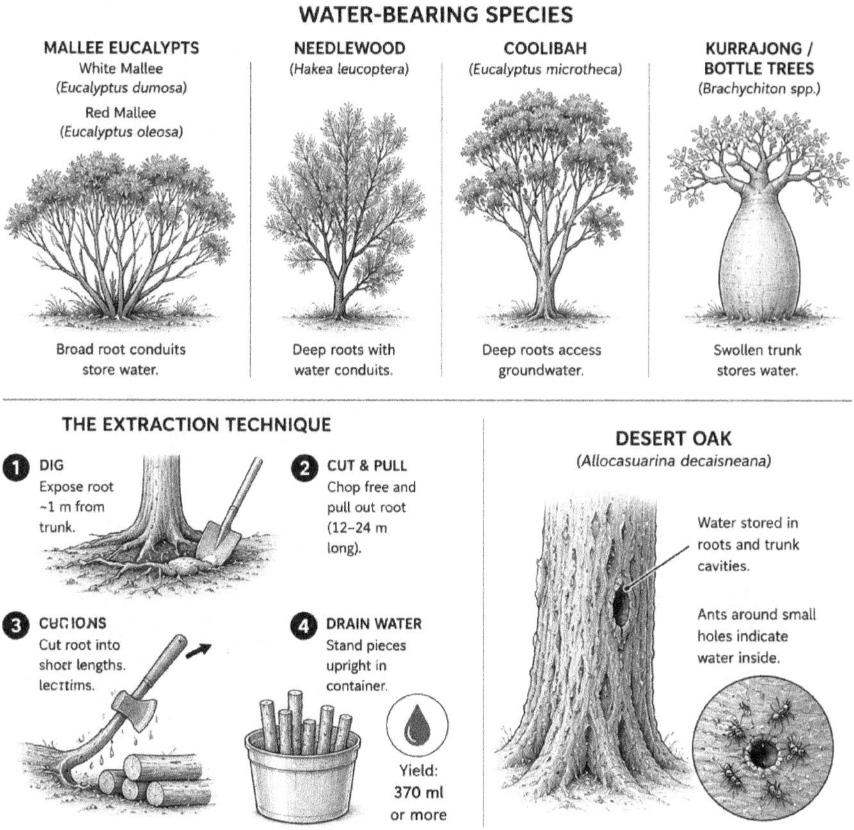

Cacti: Relevant Mainly Outside Australia

Cacti are not native to Australia, and the widely repeated survival advice to cut open a cactus for water applies to arid regions of the Americas, not to the Australian outback. Several cactus species, including some Opuntia (prickly pear) varieties, have become invasive weeds in parts of Australia and can, in principle, be used for water in the same way as their native range, but this is a minor and geographically limited option here compared to the Central Australian techniques above.

Where cacti are genuinely relevant, such as in the extreme desert isolation scenarios covered in Part VIII involving international settings, the general technique is as follows.

- Select a healthy, undamaged barrel-type cactus rather than a columnar species, since barrel cacti generally store more accessible water per unit of plant material.
- Cut off the top and mash or blend the interior pulp within the plant's own shell, taking care to avoid any spines.
- Squeeze or press the mashed pulp to extract the liquid, which can be drunk directly or filtered through cloth first to remove fibrous material.
- Avoid any cactus with visibly discoloured or unhealthy-looking flesh, and avoid species not confidently identified as non-toxic, since not all cactus species produce safe liquid.

- Cacti are not native to Australia.
- Survival advice to cut cacti for water applies to arid regions of the Americas.
- Some Opuntia (prickly pear) are invasive in parts of Australia and can, in principle, provide water, but this is limited.

GENERAL TECHNIQUE (WHERE RELEVANT)

1 SELECT CACTUS

- Choose a healthy, undamaged barrel-type cactus.
- Barrel cacti store more accessible water.

2 CUT & PREPARE

- Cut off the top.
- Mash or blend the interior pulp inside the shell.

3 EXTRACT LIQUID

- Squeeze or press mashed pulp to extract liquid.
- Drink directly or filter through cloth to remove fibrous material.

4 SAFETY FIRST

- Avoid cactus with discoloured or unhealthy-looking flesh.
- Avoid species not confidently identified as non-toxic.

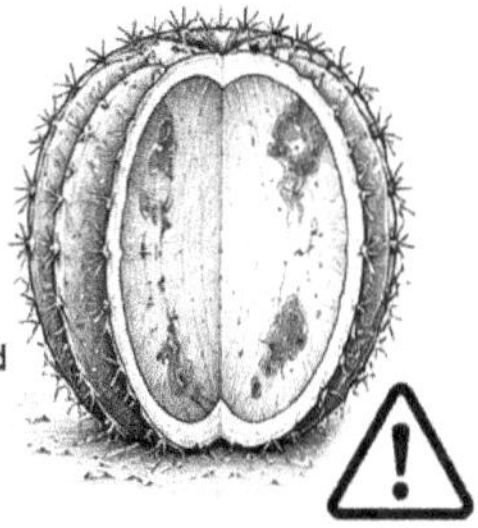

KEY POINTS

Relevant in the Americas and other international desert regions.

Barrel cacti preferred.

Liquid can be drunk or filtered.

Ensure cactus is healthy and species is non-toxic.

Coconuts: Relevant in Northern and Far North Queensland

Coconut palms grow naturally and are widely cultivated along the northern and far northern Queensland coastline and on many offshore islands in this region, making coconut water a genuinely relevant source for anyone in a survival situation in this specific part of Australia, as well as throughout the broader tropical Pacific and Southeast Asian region relevant to some of the extreme scenarios in Part VIII.

Selecting the right coconut. A green, unripe coconut generally contains more liquid, and that liquid is more palatable, than a mature brown coconut, which contains less liquid and more solid flesh.

Opening the coconut. The husk is removed first, generally by repeatedly striking the pointed end against a hard edge or spike until the fibrous husk splits and can be peeled away, exposing the harder shell beneath. The shell is then punctured at one of the three soft points near the stem end, using a knife point or a sharpened stick, to access the liquid inside without needing to fully break the shell.

Yield and safety. A single green coconut typically contains several hundred millilitres of coconut water, which is naturally sterile within an intact, undamaged nut and safe to drink directly without treatment. This makes coconut water one of the few plant-based sources in this chapter that does not require the screening and treatment set out in Parts V and VI, provided the coconut itself is intact and shows no sign of damage or spoilage before opening.

Caution with excessive consumption. Coconut water is high in potassium, and while a valuable source in the right setting, it should not be relied upon as the sole fluid source for an extended period for anyone with kidney impairment or any condition affecting potassium regulation.

Grows naturally and widely cultivated along the northern and far northern Queensland coastline and many offshore islands.

Coconut water is a relevant survival water source in this region and across the tropical Pacific and Southeast Asia.

Important for survival situations in this specific part of Australia and in scenarios covered in Part VIII.

1 SELECT THE RIGHT COCONUT

GREEN (UNRIPE)

- More liquid
- More palatable
- Best choice

VS.

BROWN (MATURE)

- Less liquid
- More solid flesh
- Not ideal

2 OPENING THE COCONUT

1. REMOVE HUSK

Strike pointed end against hard edge or spike repeatedly.

2. PEEL HUSK AWAY

Husk splits and peels away, exposing the hard shell.

3. FIND SOFT POINT

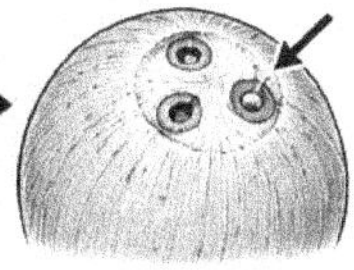

Three soft points near the stem end.

4. PUNCTURE

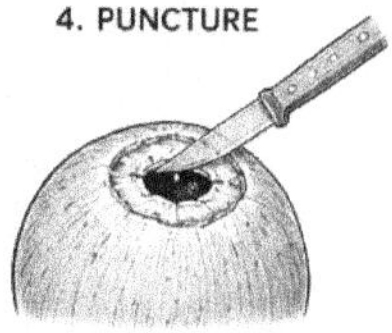

Use knife point or sharpened stick to puncture and access liquid.

3 YIELD AND SAFETY

Several hundred millilitres of coconut water per nut.

Naturally sterile inside an intact, undamaged nut.

Safe to drink directly – no treatment required.

4 CAUTION – EXCESSIVE CONSUMPTION

Coconut water is high in potassium.

Do not rely on as the sole fluid source for extended periods if you have kidney impairment or any condition affecting potassium regulation.

KEY POINTS

Common in northern and far north Queensland and many tropical regions.

Choose green coconuts for more and better tasting liquid.

Remove husk, puncture one of the three soft points to access water.

Coconut water is naturally sterile and safe to drink directly if the nut is intact and undamaged.

Dew Cups and Leaf Vessels

A further Aboriginal technique, documented particularly in relation to overnight dew collection, uses large, naturally cupped or bowl-shaped leaves, positioned to collect dew overnight, with the accumulated water then transferred each morning to a coolamon or other carrying vessel. This is a simple, low-effort supplement to the dew collection technique covered in Chapter 9, and is worth combining with it in any vegetated environment where suitably shaped leaves are available.

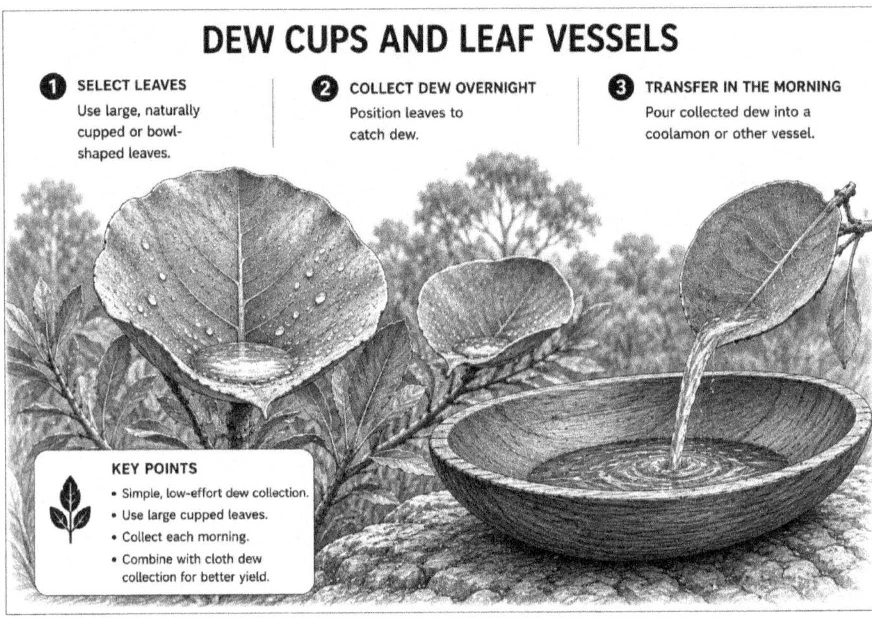

Fruit and Succulent Plant Material

Beyond dedicated water-storing roots and specialized structures, a range of fruit and succulent plant material can supplement fluid intake directly.

Ripe, unspoiled native and introduced fruit, correctly identified, provides both fluid and nutrition, and should be prioritised over extraction techniques that require tools and time where a safe, identified fruit source is directly available.

Succulent-leaved plants, correctly identified as non-toxic, can be chewed to release their fluid content, though this technique yields relatively little fluid for the effort involved compared to root extraction or fruit consumption, and should be treated as a minor supplement rather than a primary technique.

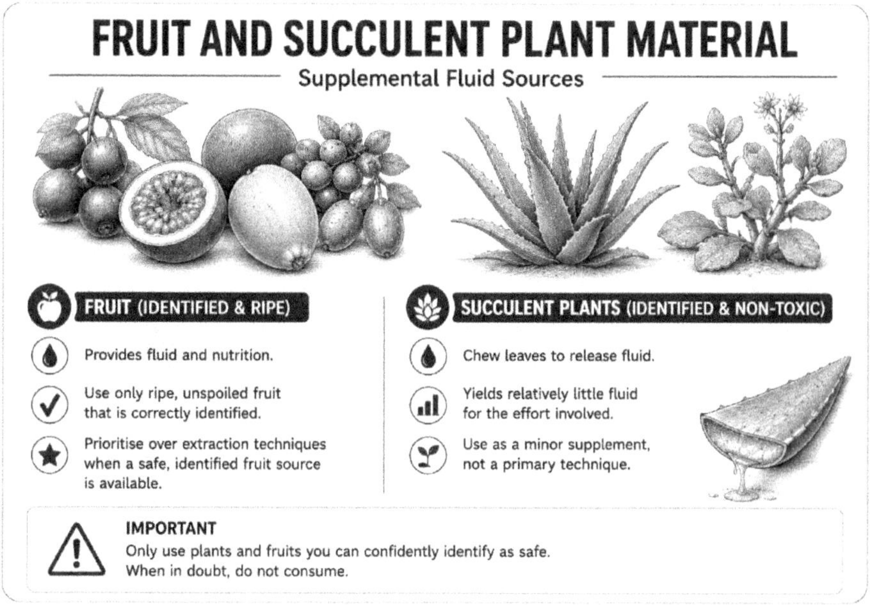

Combining Plant-Based Techniques With the Rest of This Book

Water obtained from a coconut is safe to drink directly. Water extracted from a mallee root, a cactus, or squeezed from succulent plant material carries a lower but real risk of contamination from the plant material itself, and general good practice in a survival situation is to treat any liquid extracted this way with the same caution as any other found water source, using the screening approach in Chapter 14, rather than assuming it is automatically safe simply because it came from a plant. Plant-based extraction is best understood as one further layer in the same combined strategy introduced in Chapter 9: no single technique in this section of the book is likely to meet a household's full requirement alone, but several used together, matched to the specific country and species available, meaningfully extend survival time when stored and harvested water are unavailable.

Sources referenced in this chapter

Noble, J.C. and Kimber, R.G., On the Ethno-Ecology of Mallee Root-Water, Australian National University Press

Australian Geographic, Surviving in the Desert

Queensland Government, How Aboriginal Peoples Manage Their Water Resources, resources.qld.gov.au

Australian Humanities Review, A Chant of Lost Water

CHAPTER 11

Hidden Household and Urban Sources

An ordinary Australian home holds a meaningful quantity of water outside the dedicated storage systems covered in Part II, sitting in appliances, fixtures and fittings that most households never think to check during a disruption. None of these sources should be relied upon as primary supply, and several carry specific contamination risks that make them unsuitable for drinking without treatment, or unsuitable for drinking at all. This chapter sets out what is genuinely available, what condition it is in, and how to access it safely.

The Hot Water System

A standard household hot water storage tank holds a substantial volume, typically 125 to 315 litres depending on the unit installed, and this water remains available even after mains supply has failed, provided the tank was full at the time supply was interrupted.

Turn off the power or gas supply to the unit before doing anything else, both as a safety precaution and to prevent the heating element from operating without water flowing through the system.

Turn off the cold water inlet valve at the top of the tank, if accessible, to prevent any backflow of potentially contaminated mains water into the tank once pressure returns.

Open the drain valve, typically located at the base of the tank, and run water into a clean container. Open a hot tap elsewhere in the house first, or the tank's pressure relief valve, to allow air into the system and let the water drain properly rather than forming an airlock.

Allow the water to cool before use if it has not already cooled, and treat it using the methods in Part VI if there is any doubt about the tank's internal condition, particularly in an older unit.

Older tanks, particularly those that have not been serviced or flushed regularly, can accumulate sediment at the base, and the first portion of water drained should be inspected for discolouration or particulate matter before use. This water is a genuine and often overlooked reserve, particularly valuable in the early hours of a disruption before other, larger sources have been accessed.

Toilet Cisterns, Not Toilet Bowls

The cistern, the tank that holds clean water ready for the next flush, is distinct from the bowl, and this distinction matters considerably for household water sourcing.

Cistern water is supplied directly from the mains line and, in an ordinary domestic toilet with no in-tank cleaning tablet or additive present, is the same water quality as any other tap in the house. It can be used as a genuine emergency source, ideally treated as a precaution given the fittings it has sat in contact with, but is considerably better quality than water from almost any other household source in this chapter.

Bowl water, by contrast, should never be used under any circumstances, regardless of how clean it may appear, due to the obvious sanitation risk.

Any household using an in-cistern cleaning tablet, chlorine block or blue dye additive should not use cistern water for drinking, since these products introduce chemicals into the water not intended for consumption.

Pipes and Plumbing

Water remains within household pipes even after mains pressure has failed, and can be recovered with a small amount of effort.

Open the lowest tap in the house, generally an outdoor tap or a tap in the lowest-level bathroom or laundry, and allow gravity to drain the remaining water in the pipes into a clean container.

Opening a tap at the highest point in the house simultaneously allows air to enter the system, improving the flow of water draining from the lower tap.

The volume recovered this way is modest, typically a matter of litres rather than tens of litres, but is immediately available water quality equivalent to ordinary tap water, requiring no treatment under normal circumstances.

Ice and the Freezer

A stocked freezer holds a genuine water reserve in the form of ice, ice-cube trays, and any water content in frozen food itself.

Ice melts into drinking-quality water with no treatment required, provided it was made from safe tap water and the freezer itself has remained closed and has not lost power for an extended period prior to use.

A full freezer retains its cold temperature considerably longer than a partially filled one during a power outage, which is itself a reason to keep a freezer well stocked, including with containers of plain frozen water, as part of general household preparedness rather than purely as a food storage strategy.

Once ice has fully melted and the freezer has warmed to room temperature for an extended period, any remaining liquid should be treated with the same caution as any other water that has been standing at ambient temperature for an unknown period, particularly if it has been in contact with thawing food.

Swimming Pools

A backyard swimming pool represents by far the largest volume of water available within many Australian properties, frequently tens of thousands of litres, and should be understood specifically as a bulk utility water reserve rather than a drinking water source.

Appropriate uses. Toilet flushing, cleaning, cooling, laundry, and garden irrigation are all reasonable uses for pool water during a disruption, and represent genuinely significant volume that can substantially reduce pressure on a household's drinking water reserve for every non-drinking purpose.

Why pool water is unsuitable for drinking without extensive treatment. Pool water is deliberately maintained at a chlorine concentration considerably higher than safe drinking water standards, and frequently contains additional chemicals including algaecides, pH adjusters and stabilisers not intended for ingestion. It should not be treated as an emergency drinking water source, even via boiling, unless no other option whatsoever remains available and the situation is genuinely life-threatening, in which case the specific field expedient methods covered in Part VIII for extreme scenarios should be consulted rather than treating ordinary pool water as routine drinking supply.

Access during a power outage. A pool with an electric filtration pump that has lost power remains fully usable as a stored utility reserve; only ongoing filtration and chlorination are affected, not the immediate availability of the water itself.

Waterbeds and Specialty Fixtures

Waterbeds, though less common than in previous decades, hold several hundred litres and represent a genuine, if unusual, reserve in a household that has one. This water is not suitable for drinking, since waterbed water is routinely treated with a fungicide or algaecide additive not intended for consumption, and should be treated as a utility water source only, in the same category as pool water.

Vehicles: What Is and Is Not Usable

A parked vehicle contains several fluid reservoirs, and it is important to distinguish clearly between the ones that offer any genuine emergency value and the ones that are actively dangerous.

Windscreen washer reservoir. This should never be used as a water source. Washer fluid contains detergents and, frequently, methanol or other solvents that are toxic if ingested.

Radiator and coolant system. This should never be used as a water source under any circumstances. Coolant, even when significantly diluted with water, contains ethylene glycol or propylene glycol, both of which are toxic, and ethylene glycol in particular has a sweet taste that has led to accidental poisoning, including of children and pets, in households where it has leaked or been stored carelessly. Radiator fluid should be treated as a hazardous substance requiring careful containment and disposal, never as an emergency water source.

Air conditioning condensate. A running vehicle air conditioning system produces condensate that drips from beneath the vehicle, and this water, while not guaranteed sterile, is generally low-risk and has been used as a genuine supplementary water source in documented survival situations, since it is produced by the same condensation principle as the solar stills covered in Chapter 9. It should still be treated before drinking where practical, given the unknown surfaces it has passed over.

Urban and Municipal Sources: What Not to Do

In a genuine urban disruption, several sources exist that carry legal and public health complications households should understand before considering them.

Fire hydrants. Opening a fire hydrant without explicit authorisation from the relevant fire or water authority is illegal in every Australian jurisdiction, carries a genuine risk of damaging the water pressure available to fire services during exactly the kind of event when it is most needed, and should never be attempted by a household acting independently, regardless of the apparent severity of the situation.

Ornamental fountains and water features in public or shared spaces. These are generally treated with the same category of non-potable chemical additives as a swimming pool and should be treated as a utility source only, if used at all, and only where clearly on private property the household controls.

Bringing the Household Inventory Together

A household facing a genuine water shortage benefits from working through this chapter systematically rather than reactively: draining the hot water system first, recovering standing pipe water, checking the freezer, and, where a pool is present, treating it explicitly as the largest available utility reserve for every non-drinking purpose in the house. Combined, these hidden sources frequently provide several hundred litres of immediately usable water within an ordinary home, entirely separate from the dedicated storage and harvesting systems covered earlier in this book, and often overlooked simply because they were never designed with emergency use in mind.

CHAPTER 12

Home Greywater Systems

Greywater is the lightly used wastewater generated by showers, bathroom basins, baths and laundry, and it represents one of the largest sources of usable water already present in an ordinary Australian home, most of which currently runs straight to the sewer without a second thought. This chapter sets out how to capture and use it safely, both as a routine water conservation measure and as a meaningful extension of supply during a disruption. Chapter 13 then covers the specific application of greywater to garden irrigation in more detail.

Greywater Versus Blackwater

The distinction between these two categories of household wastewater is fundamental to using either safely.

Greywater comes from showers, baths, hand basins, and laundry, with some state guidance excluding kitchen sink and dishwasher water from the greywater category due to their higher grease and food particle content. It contains soap, shampoo, small amounts of body oil and dirt, and lint, but no faecal contamination under normal circumstances.

Blackwater comes from toilets and, in most state definitions, from kitchen sinks and dishwashers, and contains faecal matter, pathogens and food waste at a level requiring full sewage treatment. Blackwater should never be captured, stored or reused by a household under any circumstances.

An average Australian household uses approximately 150 to 200 litres of water per person, per day, across all purposes, and approximately 50 to 60 percent of this qualifies as greywater under the definition above. This represents a genuinely significant volume: for a household of four, potentially 300 to 480 litres per day of water that currently goes straight to the sewer and could, with the right system, be captured for secondary use instead.

Simple, Zero-Cost Capture Methods

The most accessible entry point requires no plumbing modification and no expenditure at all.

The shower bucket. Placing a clean bucket in the shower to catch water while bathing is the simplest possible greywater system. A typical shower runs at 8 to 12 litres per minute, and a three-minute shower using this method recovers the majority of that flow for reuse.

The basin plug method. Using the basin plug during handwashing and tooth brushing, rather than allowing water to run directly down the drain, and transferring the collected water to a bucket, recovers a further, smaller but genuinely useful volume across the day.

Laundry rinse capture. Diverting the final rinse cycle of a washing machine into a container, either manually or via a simple redirected outlet hose, recovers water that is typically cleaner than the initial wash cycle, having already had most detergent removed by the earlier rinse.

Handling captured water. Greywater captured this way should be used within 24 hours to avoid bacterial growth, particularly in warm weather, since it contains organic material such as soap residue, skin cells and hair that supports microbial growth once it stands for an extended period. It should never be used on food plants intended for raw consumption, and should be applied to soil rather than sprayed in a way that creates fine mist, to reduce any risk of inhalation exposure.

Installed Diversion Systems

For a household planning for extended greywater reuse, particularly during longer disruptions or as an ongoing water-saving measure, a simple gravity-fed diversion system redirects shower and bath drainage to a storage container or directly to subsurface garden irrigation, rather than to the sewer.

Greywater diversion devices (GDDs) capture greywater and direct it, generally without treatment, straight to garden irrigation, typically via a subsurface or below-mulch distribution system that avoids direct human contact with the water.

Greywater treatment systems (GTSs) apply some level of filtration or biological treatment before distributing the water, allowing broader reuse, including in some cases surface irrigation or non-potable indoor uses such as toilet flushing, depending on the treatment standard achieved and the specific approval granted.

Both categories of system are covered in Australia by AS/NZS 1547 for on-site wastewater management, and by the sector's standard reference document, the Urban Greywater Installation Handbook for Single Households (HB 326-2008), which sets out best-practice control measures for both treated and untreated greywater reuse.

Regulatory Requirements

Greywater reuse is regulated at state level in Australia, and requirements vary in their detail, though the underlying principles are broadly consistent.

Minimum setback distances from waterways, boundaries, and in-ground water sources such as bores must be observed for any subsurface irrigation system, to prevent surface or groundwater contamination.

Prohibition on food plants intended for raw consumption. Untreated greywater should not be applied to plants such as lettuce, herbs or other produce eaten raw and unwashed, due to the residual bacterial load it carries. Treated greywater from an approved GTS may have different, less restrictive conditions attached, depending on the treatment standard achieved.

Licensed plumber requirement. Any modification to household plumbing to install a diversion system, including redirecting a drain, must be carried out by a licensed plumber in every Australian state, and the installation is typically regulated under the AS/NZS 3500 plumbing standards series.

Council approval. Many local councils require notification or approval before a greywater diversion system is installed, even where the state-level regulatory framework does not require a specific licence for the greywater use itself, and confirming local requirements before installation avoids a costly retrofit later.

Water Conservation Practices That Extend Any Reserve

Reducing the amount of water a household consumes in the first place is the least discussed but most immediately actionable water security measure available, and several straightforward changes meaningfully extend any stored, harvested or captured reserve during a genuine scarcity event.

Sponge bathing instead of showering reduces water use for washing from 40 to 60 litres to as little as 2 to 4 litres per person, per wash.

Reusing cooking water. Water used to cook pasta or rice, once cooled, can serve as a base for soup or stock in a subsequent meal, rather than being discarded.

Sequential dishwater use. Washing dishes in a single basin of water, then reusing the same water for a second, lower-priority cleaning task such as wiping down surfaces, extracts more use from the same volume before disposal.

Reusing drinking glasses. Where hygiene conditions allow, the same household member reusing their own drinking glass across a day, rather than using a fresh glass each time, reduces the washing water required.

Eliminating discretionary water use entirely during a genuine scarcity event, including car washing, hosing hard surfaces and watering ornamental, non-food garden beds, removes uses that add no genuine household value during a disruption.

Building a Household Greywater Plan

Start with the zero-cost shower bucket and basin plug methods, which require no installation and no regulatory approval, and are appropriate for every household regardless of tenure or budget.

For a household in a fixed home with a longer-term view, investigate a properly installed diversion system, engaging a licensed plumber and confirming council and state requirements before work begins.

Establish clear household rules for what greywater is used for, such as irrigation, toilet flushing and cleaning, and what it is never used for, such as drinking, raw food plants or direct skin contact for an extended period, and make sure every household member understands the distinction between greywater and the fully treated water in the household's dedicated drinking reserve.

Treat greywater capture as a genuine extension of the household's total water security, worth factoring into the planning calculations from Chapter 2, rather than as a minor, incidental saving.

Sources referenced in this chapter

Standards Australia, Urban Greywater Installation Handbook for Single Households, HB 326-2008

AS/NZS 1547:2012, On-site domestic wastewater management

AS/NZS 3500 series, Plumbing and drainage

Department of Health Western Australia, A Compilation of Australian Standards on Water Holding Tanks (greywater provisions)

CHAPTER 13

Greywater for Irrigation and the Garden

Garden irrigation is the single largest and most valuable ongoing use for household greywater, and it is also the use most likely to go wrong if approached carelessly. Greywater is not equivalent to rainwater or treated mains water from a garden's perspective: it carries dissolved soap, detergent residue and, depending on the products used in the household, sodium, phosphorus and boron at levels that can genuinely damage soil structure and plant health over time if applied without care. This chapter sets out how to use greywater to keep a garden alive through an extended water restriction or supply disruption without causing that damage.

Choosing Detergents and Soaps With Irrigation in Mind

The single most important decision affecting whether a greywater irrigation system helps or harms a garden over time is made at the supermarket, not in the garden itself.

Sodium. Many detergents and soaps use sodium-based surfactants, and repeated application of high-sodium greywater degrades soil structure over time, particularly in clay soils, by displacing other minerals plants need and reducing the soil's ability to drain and aerate properly.

Boron. Boron, present in some laundry products, particularly certain whitening and stain-removal formulations, is toxic to a wide range of plants at concentrations far lower than would concern human health, and is one of the most common causes of greywater-related plant damage in Australian gardens.

Bleach and strong disinfectants. Any laundry load washed with bleach or a strong chlorine-based disinfectant should not have its wastewater directed to garden irrigation; divert this specific load to the sewer as normal, and reserve greywater capture for ordinary wash loads using a garden-safe product.

Phosphorus. Moderate phosphorus levels, common in many household detergents, are generally tolerated by gardens in the way garden fertilisers already supply phosphorus, but very high levels can still cause nutrient imbalance over an extended period of exclusive greywater irrigation.

Choosing a garden-safe product. A liquid, low-sodium, low-phosphate, boron-free laundry detergent and body wash, widely available in Australian supermarkets and often explicitly marketed as greywater-safe or garden-safe, substantially reduces the risk of long-term soil and plant damage from an ongoing greywater irrigation system.

Which Plants Tolerate Greywater Well

Plant tolerance for greywater varies considerably, and matching the right plants to the greywater zone of a garden meaningfully reduces the risk of damage.

Generally tolerant: established citrus trees, roses, most Australian native shrubs once established, hardy ornamental grasses, and most fruit trees once past their first year or two in the ground.

Generally sensitive: azaleas, camellias, ferns, and many acid-loving ornamentals, which are particularly vulnerable to the change in soil pH and salinity that regular greywater application can cause.

Newly planted and seedling plants of almost any species are more vulnerable than established plants, since their root systems have not yet developed the resilience that comes with maturity, and greywater irrigation is best reserved for established garden zones rather than a new seedling bed.

Soil Type and Application Method

The same volume of greywater behaves very differently depending on the soil it is applied to, and the application method used.

Sandy soils drain quickly and are less prone to the salt accumulation that affects clay soils, but also hold moisture and nutrients for a shorter period, meaning greywater needs to be applied more frequently in smaller amounts to be effective.

Clay soils drain slowly and are considerably more prone to long-term salt and sodium accumulation from repeated greywater application, making the detergent selection guidance above particularly important for a garden on clay.

Subsurface or mulch-basin application, where greywater is directed beneath a layer of mulch or directly into the root zone via a simple trench or basin around each plant, rather than sprayed across the surface, is the safest general approach. It reduces human contact with the water, reduces odour, and reduces the evaporative loss that surface application experiences in hot Australian conditions.

Never spray greywater as a fine mist, both because of the inhalation risk from any aerosolised bacterial content and because misting is a comparatively inefficient way to get water to the root zone where the plant can actually use it.

Rotation and Rest Periods

A garden irrigated exclusively with greywater over an extended period benefits from a rotation system, rather than continuous application to the same beds indefinitely.

Divide the garden into at least two irrigation zones where practical.

Direct greywater to one zone for a period, while the other zone rests and receives only rainwater or a reduced application.

Rotate between zones on a period of several weeks to a few months, depending on soil type, allowing each zone time to process accumulated nutrients and salts through normal soil biological activity before the next application cycle begins.

Where a household has only a single garden bed or a very small yard, an occasional flush of the same bed with plain rainwater or a larger-than-usual volume of clean water helps leach accumulated salts below the root zone and reduces the risk of long-term build-up.

Greywater and Food Gardens

Applying greywater to a productive food garden requires more caution than applying it to ornamental plants, because the risk profile changes once the produce itself may be eaten raw and unwashed.

Fruit trees are generally suitable for greywater irrigation at the root zone, provided the water does not contact the fruit itself directly, since the fruit is typically elevated above the irrigation zone and any residue on the fruit's surface can, in any case, be washed off before eating.

Root vegetables and leafy greens eaten raw, including lettuce, spinach, herbs and carrots, should not be irrigated with untreated greywater, since these crops are the most likely to be eaten with minimal washing and the most likely to have direct contact between the edible portion and the irrigation water or soil.

Cooked vegetables, where the final preparation step involves genuine heat treatment, carry a lower risk profile, but growing them with mains water, rainwater or well-treated greywater from an approved GTS remains the more conservative choice for any household uncertain about the specific risk involved.

Where any doubt exists about the safety of a specific crop and greywater combination, the safer default is to reserve greywater for the ornamental and non-food sections of the garden and use rainwater, harvested water or mains supply, in that order of priority, for anything intended to be eaten raw.

Recognising Greywater Damage Early

A garden receiving greywater irrigation should be monitored periodically for early signs of damage, since correcting a problem early is considerably easier than reversing established soil damage.

Leaf tip and margin burn, appearing as browning or dead tissue at the very edges and tips of leaves, is a classic symptom of salt or sodium accumulation in the soil.

General decline without an obvious cause, including reduced flowering, stunted growth or leaf drop in a plant that was previously healthy, warrants a review of which detergents have been used and whether a rest period or a flush with clean water is needed.

Visible white crust on the soil surface, particularly in clay soils in warmer weather, is a direct sign of salt accumulation at the surface as water evaporates and leaves dissolved salts behind, and is a clear signal to reduce greywater application to that specific area and increase the frequency of clean water flushing.

Sources referenced in this chapter

Standards Australia, Urban Greywater Installation Handbook for Single Households, HB 326-2008

Australian state environment and health department guidance on greywater use in home gardens

General horticultural guidance on soil salinity and sodium accumulation from irrigation water

CHAPTER 14

Screening Raw Water

Before any water located, harvested or extracted using the methods in Parts II, III and IV is treated using the methods in Part VI, it should be assessed. Screening is not a substitute for treatment, and nothing in this chapter should be read as a method for confirming water is safe to drink without further action. Its purpose is narrower and still essential: to identify sources that should be avoided altogether regardless of treatment, to identify sources that need a higher level of treatment than others, and to build the habit of assessing water deliberately rather than assuming safety based on appearance alone.

The Fundamental Rule

No natural water source in Australia should be assumed safe without treatment. Clear, fast-flowing, visually appealing water from a remote stream can carry Giardia cysts, Cryptosporidium oocysts, agricultural runoff or livestock faecal contamination that are completely invisible to the naked eye. Aesthetic quality has no reliable relationship to microbiological safety, and this rule applies regardless of how remote the source, how clear the water, or how confident the observer feels about the surrounding land use.

Visual and Sensory Assessment

A brief, structured visual assessment, conducted before collection, helps identify the sources that carry the highest risk and should be avoided in favour of an alternative wherever one exists.

Colour. Water with a distinct brown, yellow, green or grey tint indicates dissolved organic material, algae, sediment or, in some cases, chemical contamination. Some tannin-stained water from healthy bushland catchments is a naturally occurring, relatively low-risk brown colour, but this should not be assumed without further consideration of the surrounding land use.

Clarity and turbidity. Cloudy or turbid water carries suspended sediment that can shield pathogens from chemical and ultraviolet treatment, and should generally be allowed to settle, or pre-filtered through cloth, before any treatment method is applied.

Smell. A musty, earthy, sulphurous or otherwise off odour is a warning sign, though the absence of odour does not confirm safety, since many dangerous contaminants, including Giardia and Cryptosporidium, are entirely odourless.

Surface film or scum. A greenish, bluish, brown or reddish scum or paint-like coating on the surface indicates possible blue-green algal bloom, discussed further below, and water showing this sign should be avoided entirely rather than treated.

Surrounding evidence. Dead animals in or near the water, heavy livestock traffic at the water's edge, nearby agricultural land, road crossings or any visible industrial activity upstream all raise the likely contamination level of a source, even where the water itself appears clear.

Blue-Green Algal Bloom: A Specific, Serious Hazard

Blue-green algae, more accurately cyanobacteria, form blooms in warm, nutrient-rich, typically still water across much of Australia, particularly in farm dams, slow rivers and lakes during summer. This hazard deserves separate treatment because the standard advice given for every other biological contaminant in this book, namely that boiling and standard filtration render water safe, does not apply here.

Affected water shows a greenish, bluish, brown or reddish scum, paint-like coating, or streaking on the surface, frequently with a distinctive musty or earthy smell.

The cyanotoxins these organisms produce are not removed by boiling and are not reliably removed by standard portable water filters, since the toxins are dissolved rather than particulate.

Exposure can cause liver damage, nervous system effects, skin irritation, and has caused death in livestock and, at sufficient exposure, in humans.

Water showing any sign of algal bloom must be avoided entirely for drinking, cooking, or animal use. There is no treatment method covered in this book that reliably neutralises cyanotoxins in the field, and the only safe response is to use an alternative, unaffected source.

Field Testing for Biological Contamination

Beyond visual assessment, simple field test kits provide a further, more objective layer of screening, though they should be understood as indicative rather than definitive.

Coliform bacteria test strips and tube tests, available from water quality testing suppliers, detect the presence of coliform bacteria, the standard indicator organism for faecal contamination, typically within 24 to 48 hours of incubation.

A positive result reliably signals that treatment is required and should never be ignored.

A negative result provides reasonable, though not absolute, reassurance for a clear water source in a low-risk environment, but does not confirm the complete absence of viruses, protozoa or chemical contamination, which these tests are not designed to detect.

These kits are a genuinely useful addition to a household or field kit, but they introduce a delay of a day or more before a result is available, meaning they suit planning and risk assessment for a semi-permanent camp or property far better than an urgent, immediate drinking decision.

Field Testing for Chemical Contamination

Chemical contamination is harder to screen for in the field than biological contamination, and this limitation should be understood clearly rather than papered over.

Simple chemical test strips can indicate pH, chlorine residual, and, in some more comprehensive kits, the presence of specific heavy metals or common agricultural chemicals, but no field kit reliably screens for the full range of possible chemical contaminants in a flood-affected, industrial or agricultural setting.

In any flood-affected area, all water from affected sources should be treated as potentially chemically contaminated, since floodwater collects agricultural chemicals, fuel, sewage and industrial discharge along its path, until official guidance from the relevant water authority confirms otherwise.

Similarly, water near a known chemical spill, industrial accident, or any deliberate contamination event covered further in Part VIII should not be treated using the biological methods in Part VI alone, since none of those methods reliably remove dissolved chemical contamination.

When to Trust an Official Boil-Water Notice

Following a flood, an infrastructure failure, or any event that may have compromised mains water quality, the relevant water authority may issue a boil-water notice, instructing residents to boil tap water before drinking, cooking or brushing teeth until the notice is formally lifted.

These notices should be followed for their full duration, and water should not be assumed safe again simply because it looks and smells normal before the notice has been officially withdrawn.

Water used for infant formula, for anyone immunocompromised, and for any medical use should be treated with particular caution during and immediately after a boil-water notice period, even after the notice is lifted, if there is any reason to suspect the household's specific plumbing may have been affected differently from the wider network.

A Practical Screening Decision Framework

For any water source that is not already known and trusted, work through the following sequence before deciding how, or whether, to proceed.

Conduct the visual and sensory assessment set out earlier in this chapter. If any sign of algal bloom is present, stop and find an alternative source; no further step in this framework applies.

Note the surrounding land use and any visible contamination risk, including livestock access, agricultural land, road crossings and industrial or urban proximity.

Where time and equipment allow, run a field coliform test to inform, rather than replace, the judgement formed in the first two steps.

Select a treatment approach from Part VI matched to the assessed risk level: a single treatment method for a relatively low-risk source such as a clean, flowing highland stream with no upstream human activity, and the full multi-barrier approach set out in Chapter 19 for any source with an elevated or uncertain risk profile.

Where any doubt remains after working through the above steps, apply the more thorough treatment option available rather than the faster one. The cost of additional treatment is a few minutes of time. The cost of inadequate treatment is a waterborne illness at precisely the moment the household can least afford to manage one.

Sources referenced in this chapter

National Park Service (US), 2 Ways to Purify Water, nps.gov

Queensland and New South Wales health department guidance on blue-green algal bloom

Australian Drinking Water Guidelines, boil-water advisory principles

Bushwalking Manual Australia, water treatment options, bushwalkingmanual.org.au

CHAPTER 15

Understanding Contaminants

Every treatment method covered in Part VI works by exploiting a specific physical or chemical vulnerability of a specific category of contaminant. Boiling kills biological pathogens through heat but does nothing to dissolved chemicals. A 0.2-micron filter physically strains out bacteria and protozoa but passes viruses through unaffected. Chlorine dioxide addresses the full biological spectrum but has no effect on dissolved heavy metals. Understanding what each contaminant actually is, its size, its structure and its specific weaknesses, makes it possible to choose the right treatment for a given source with confidence, rather than applying a method and hoping it happens to work. This chapter sets out the four broad categories of water contaminant relevant to Australian conditions: bacteria, viruses, protozoa and chemical contamination, along with the specific case of cyanotoxins introduced in Chapter 14.

Bacteria

Bacteria are single-celled microorganisms, typically ranging from 0.5 to 10 microns in size, that reproduce independently once inside a suitable host or environment.

Escherichia coli (E. coli). The standard indicator organism used in water testing, most strains of E. coli are not themselves dangerous, but their presence indicates faecal contamination and the likely presence of more dangerous organisms alongside it. Certain pathogenic strains cause severe gastrointestinal illness directly.

Salmonella. Causes salmonellosis, with symptoms including diarrhoea, fever and abdominal cramps, typically appearing within 12 to 72 hours of exposure and commonly associated with contamination from animal or human faecal matter reaching a water source.

Campylobacter. One of the most common causes of bacterial gastroenteritis associated with contaminated water in Australia, frequently linked to livestock and poultry faecal contamination reaching agricultural waterways.

Vulnerability to treatment. Bacteria are reliably killed by boiling, by chemical disinfection with chlorine dioxide or iodine at standard dosing, and are physically removed by any filter rated at 0.2 microns or finer, since bacteria fall entirely within the size range these filters are designed to exclude.

Protozoa

Protozoa are single-celled organisms, generally larger than bacteria, that form environmentally resistant cysts or oocysts capable of surviving outside a host for extended periods, including in cold water.

Giardia (causing giardiasis). Giardia cysts range from approximately 8 to 12 microns, are common in Australian waterways affected by livestock, wildlife or human faecal contamination, and cause diarrhoea, cramping, bloating and fatigue that can persist for weeks if untreated.

Cryptosporidium (causing cryptosporidiosis). Cryptosporidium oocysts are smaller than Giardia cysts, at approximately 4 to 6 microns, and are the most heat-resistant and chemically resistant common waterborne pathogen relevant to Australian conditions. Cryptosporidium is markedly more resistant to standard chlorine and iodine dosing than bacteria or Giardia, requiring either a longer contact time, a higher dose, or reliance on filtration or boiling instead.

Vulnerability to treatment. Both organisms are reliably killed by boiling, reliably removed by any filter rated at 0.2 microns or finer given their larger size relative to bacteria, and require closer attention to chemical treatment dosing and contact time than bacteria do, particularly for Cryptosporidium.

Viruses

Viruses are considerably smaller than bacteria or protozoa, typically ranging from 0.02 to 0.3 microns, and this size difference has a direct and important practical consequence for water treatment.

Hepatitis A and **norovirus** are the viruses of greatest practical concern in water contaminated by human sewage, both capable of causing serious illness from a very small infective dose.

Rotavirus is a further common cause of viral gastroenteritis associated with contaminated water, particularly affecting infants and young children more severely than adults.

Vulnerability to treatment. Viruses are reliably killed by boiling and are reliably inactivated by chemical disinfection at standard chlorine dioxide dosing. They are not reliably removed by a standard 0.2-micron mechanical filter, since viruses are smaller than the filter's pore size and pass through largely unaffected. This is the single most important reason a mechanical filter should never be treated as a complete treatment solution on its own, particularly for any water source with a realistic risk of human sewage contamination, and is the basis for the multi-barrier approach covered in Chapter 19.

Chemical Contaminants

Chemical contamination behaves entirely differently from biological contamination and requires an entirely different treatment approach, since none of the biological treatment methods in this book (boiling, standard filtration, or chlorine-based disinfection) reliably remove dissolved chemical contaminants.

Heavy metals, including lead, arsenic and cadmium, can enter water from old plumbing and roof materials (as covered for rainwater tanks in Chapter 6), from mining and industrial activity, and from some naturally occurring geological sources in groundwater. Removal generally requires activated carbon filtration, reverse osmosis, or distillation, rather than the biological treatment methods covered in Chapters 16 to 18.

Agricultural chemicals, including pesticides and herbicides, can reach both surface water and groundwater in agricultural areas, particularly after heavy rain, and are a specific concern for any water source downstream of, or adjacent to, farmland. Activated carbon filtration provides meaningful reduction of many of these compounds, though not complete removal, and distillation is the most reliable field-achievable method for full removal.

Fuel and hydrocarbon contamination can occur near roads, fuel storage, industrial sites, and following a flood event that has disturbed fuel storage tanks or vehicles. This contamination is often detectable by smell and, in more severe cases, by a visible sheen on the water surface, and affected water should not be used even after boiling, since boiling can concentrate rather than remove some volatile compounds.

Nitrates, associated with agricultural fertiliser runoff and, in some cases, with septic system contamination of groundwater, pose a specific risk to infants, since high nitrate levels can interfere with an infant's oxygen-carrying capacity in the blood, a condition known as methaemoglobinaemia. Any bore or well water intended for infant formula preparation in an agricultural area should ideally be tested for nitrate levels specifically, given this distinct and serious risk.

Cyanotoxins: The Exception to Every Rule

As introduced in Chapter 14, cyanobacteria (blue-green algae) produce toxins, principally microcystins and, in some bloom types, saxitoxins, that behave unlike every other contaminant covered in this chapter. They are not removed by boiling, since they are heat-stable, and they are not reliably removed by standard mechanical filtration or standard chemical disinfection, since they are dissolved rather than particulate and are not effectively neutralized by chlorine dioxide at typical field dosing. Water affected by a visible algal bloom, as described in Chapter 14, must be avoided entirely rather than treated, because no treatment method covered in Part VI reliably addresses this specific hazard in field conditions.

Matching Treatment to Contaminant: A Summary Reference

The table below summarizes which treatment categories, covered in full in Chapters 16 through 19, are effective against each contaminant category covered in this chapter.

Bacteria: effectively addressed by boiling, standard filtration (0.2 microns or finer), and chemical disinfection.

Protozoa (Giardia, Cryptosporidium): effectively addressed by boiling and standard filtration; addressed by chemical disinfection with correct, and in the case of Cryptosporidium, extended dosing.

Viruses: effectively addressed by boiling and chemical disinfection; not reliably addressed by standard mechanical filtration alone.

Heavy metals and dissolved chemicals: not addressed by boiling, standard filtration, or standard chemical disinfection; requires activated carbon filtration, reverse osmosis or distillation.

Cyanotoxins: not reliably addressed by any field treatment method covered in this book; avoidance of the affected source is the only safe response.

UNDERSTANDING CONTAMINANTS

Four broad categories of water contaminant relevant to Australian conditions, plus the specific case of cyanotoxins.

1. BACTERIA

Single-celled microorganisms, typically 0.5 – 10 microns, that reproduce once inside a suitable host or environment.

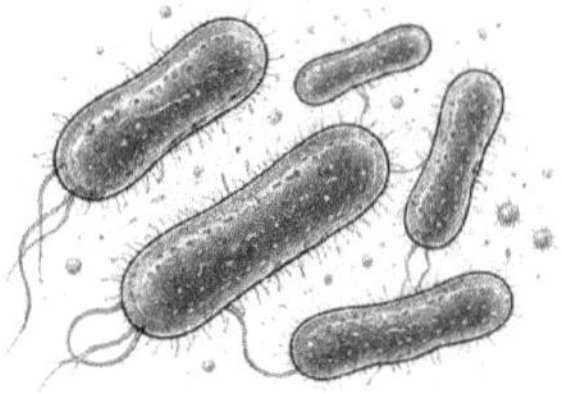

EXAMPLES

- *Escherichia coli (E. coli)*
- *Salmonella*
- *Campylobacter*

VULNERABLE TO

Boiling | Chemical disinfection | 0.2 µm filtration

2. PROTOZOA

Single-celled organisms, larger than bacteria, that form resistant cysts or oocysts and can survive in the environment for long periods.

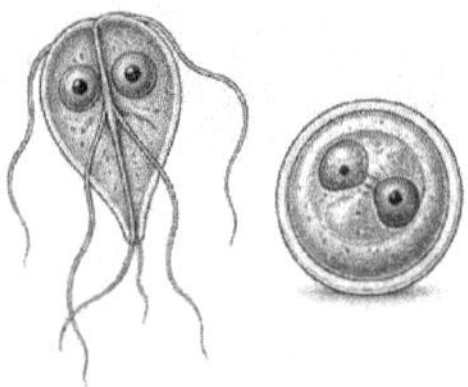

EXAMPLES

- *Giardia* (giardiasis)
- *Cryptosporidium*

VULNERABLE TO

Boiling | Chemical disinfection (extended dose) | 0.2 µm filtration

3. VIRUSES

Much smaller than bacteria or protozoa, typically 0.02 – 0.3 microns.

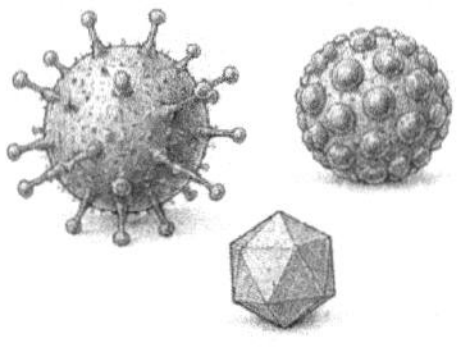

EXAMPLES

- Hepatitis A
- Norovirus
- Rotavirus

VULNERABLE TO

Boiling | Chemical disinfection | ✗ Not reliably removed by 0.2 µm filtration

4. CHEMICAL CONTAMINANTS

Dissolved or non-biological contaminants. Require different treatment approaches.

EXAMPLES

- Heavy metals (lead, arsenic, cadmium)
- Agricultural chemicals (pesticides, herbicides)
- Fuel & hydrocarbons
- Nitrates (fertiliser or septic contamination)

TREATMENT REQUIRED

Activated carbon | Reverse osmosis | Distillation

5. CYANOTOXINS: THE EXCEPTION TO EVERY RULE

Toxins produced by cyanobacteria (blue-green algae). Not removed by boiling, standard filtration, or chemical disinfection.
Avoid affected water entirely.

⚠ Not reliably addressed by any field treatment method in this book. Avoid the source.

MATCHING TREATMENT TO CONTAMINANT: SUMMARY REFERENCE

CONTAMINANT	BOILING	0.2 µm FILTRATION (OR FINER)	CHEMICAL DISINFECTION (STANDARD DOSE)	OTHER TREATMENT REQUIRED
Bacteria	✓	✓	✓	–
Protozoa (*Giardia, Cryptosporidium*)	✓	✓	✓ (extended dose)	–
Viruses	✓	✗	✓	–
Heavy metals & dissolved chemicals	✗	✗	✗	Activated carbon, RO, or distillation
Cyanotoxins (microcystins, saxitoxins)	✗	✗	✗	Avoid source

✓ Effective ✗ Not effective / Not reliable – Not applicable

This summary is the reason Part VI does not present a single "correct" treatment method, but instead builds toward the multi-barrier approach in Chapter 19: no single method covers every contaminant category, and a source of uncertain or mixed risk, which describes the great majority of natural water sources encountered in a genuine emergency, requires more than one treatment method applied in sequence to achieve a genuine, comprehensive standard of safety.

Sources referenced in this chapter

Australian Drinking Water Guidelines (NHMRC), pathogen and chemical contaminant profiles

Queensland Health and NSW Health, blue-green algal bloom and cyanotoxin guidance

World Extreme Medicine, Purifying Wild Water, worldextrememedicine.com

General public health guidance on nitrate contamination and infant methaemoglobinaemia risk

CHAPTER 16

Boiling

Boiling is the oldest, most reliable and most universally understood water treatment method available. It remains the definitive treatment for biological contamination because it kills every category of biological pathogen covered in Chapter 15 without exception, requires nothing beyond a heat source and a container, and provides an unmistakable, visually confirmable endpoint. Its limitations are equally important to understand: it does nothing to remove chemical contamination, and in some cases it can make chemical contamination worse rather than better.

The Physics of Pathogen Destruction

The temperatures required to kill waterborne pathogens are well established and considerably lower than the 100 degrees Celsius reached at a rolling boil at sea level, which is why boiling provides such a substantial safety margin.

All common waterborne bacterial pathogens are killed at temperatures below 70 degrees Celsius with sufficient exposure time.

Giardia cysts are killed at 70 degrees Celsius.

Cryptosporidium oocysts, the most heat-resistant common waterborne pathogen relevant to Australian conditions, are killed at 74 degrees Celsius held for one minute.

All waterborne viruses relevant to this book are killed well below the boiling point.

At the 100 degrees Celsius reached at a rolling boil at sea level, destruction of every organism above is essentially instantaneous rather than requiring the full stated exposure time, which is why the standard field guidance of a one-minute rolling boil provides substantial overkill rather than a bare minimum.

The Standard Boiling Protocol

- Bring the water to a rolling boil, meaning large, vigorous bubbles breaking the surface continuously, not merely small bubbles forming at the base of the container.
- Maintain the rolling boil for one minute at sea level.
- At altitudes above 2,000 metres, where water boils at a lower temperature due to reduced atmospheric pressure, extend the boiling time to three minutes to maintain the same safety margin. This adjustment is relevant to anyone in Australia's higher alpine country, and becomes more broadly relevant for the high-altitude scenarios covered in Part VIII.
- Allow the water to cool in a covered container, to prevent recontamination from airborne dust or insects settling on the open surface while it cools.

Boiling Water That Contains Visible Sediment or Contamination

Water drawn from a natural source, a hand-dug well, or any source carrying visible particulate matter benefits from a specific two-stage approach that improves both the safety and the palatability of the final result.

Where possible, allow the water to stand undisturbed before boiling, so heavier sediment settles to the bottom of the container under gravity.

Bring the water to a rolling boil for the full duration set out above.

Allow the boiled water to cool with the container undisturbed, so any remaining sediment continues to settle to the bottom rather than being stirred back into suspension.

Carefully pour or ladle off the upper three-quarters of the water into a clean, separate container for drinking, leaving the bottom quarter, along with the settled sediment it contains, behind in the original vessel.

Discard the remaining sediment-laden water, or reserve it for a non-drinking use such as cleaning, rather than attempting to filter or otherwise recover it for consumption.

This technique does not replace the temperature and time requirements set out above, both of which must still be met in full. It addresses a separate problem: boiling kills the organisms present in the water, but it does not remove settled particulate matter, and consuming the sediment layer directly, even after boiling, is both unpleasant and unnecessary when the same container's upper water is available and cleaner. This approach is particularly worth using for water drawn from a hand-dug emergency well, as covered in Chapter 7, or from a visibly turbid natural source, where sediment load is highest.

Fuel-Efficient Heat Treatment

Boiling requires fuel, and fuel is frequently a limiting resource in exactly the kind of extended disruption this book is concerned with. Several techniques reduce the fuel required to reach a safe treatment standard.

Pasteurisation rather than a full boil. Because the pathogen-killing temperatures set out earlier in this chapter are considerably below 100 degrees Celsius, heating water to approximately 65 degrees Celsius and holding it at that temperature for several minutes achieves the same biological safety outcome as a full rolling boil, using meaningfully less fuel. A Water Pasteurisation Indicator (WAPI), a small, inexpensive wax-filled device used in international water treatment programs, changes state at the correct pasteurisation temperature and removes the guesswork from this approach, allowing a household to stop heating as soon as the indicator confirms the target temperature has been reached and held, rather than continuing on to a full, more fuel-intensive boil.

Retained-heat cooking. Bringing water to a boil, then transferring the sealed pot into an insulated container, such as a box lined with towels or sleeping bags, allows the residual heat to continue treating the water for an extended period without further fuel input, since the actual temperature and time requirements are modest relative to the heat retained in a large volume of near-boiling water.

Solar ovens and reflective concentrators, covered in more detail elsewhere in household preparedness literature, can bring water to a full boil using only solar energy in suitable Australian conditions, entirely eliminating fuel consumption for water treatment on a sunny day.

Maximise the value of each boiling session by treating the largest practical volume in a single session, rather than repeatedly heating small batches, since a large pot loses proportionally less energy to its surroundings than several smaller batches processed separately.

Improving Taste

Some people notice a slightly flat or metallic taste in boiled water, caused by the reduction in dissolved oxygen during heating. This has no bearing on safety and can be improved, where desired, by pouring the cooled water back and forth between two clean containers several times to reintroduce air, or by adding a small pinch of salt.

BOILING: THE DEFINITIVE WATER TREATMENT

Kills all biological pathogens. Simple. Reliable. Proven.

1. WHY BOILING WORKS

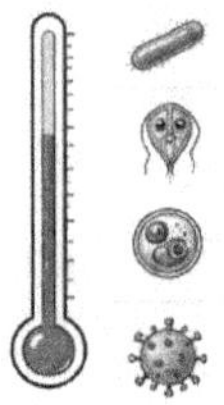

- Bacteria – killed below 70°C
- Giardia cysts – killed at 70°C
- Cryptosporidium oocysts – killed at 74°C held for 1 minute
- Viruses – killed well below boiling point

At 100°C (rolling boil), destruction is essentially instantaneous.

2. THE STANDARD BOILING PROTOCOL

1. Bring water to a rolling boil.
2. Maintain a rolling boil for 1 minute at sea level.
3. Above 2,000 m altitude, extend boiling time to 3 minutes.
4. Allow water to cool in a covered container.

3. WATER WITH VISIBLE SEDIMENT – TWO-STAGE APPROACH

1. Let water stand. Sediment settles.
2. Boil for the full time.
3. Cool undisturbed. Sediment continues to settle.
4. Carefully pour off upper 3/4 into clean container.
5. Leave bottom 1/4 and sediment behind.
6. Discard sediment-laden water or use for non-drinking.

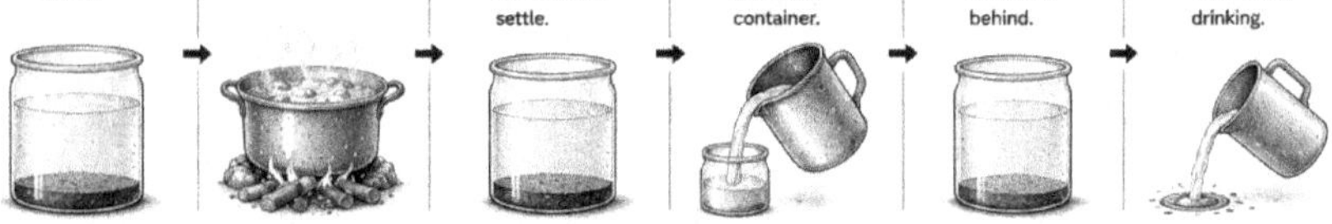

4. FUEL-EFFICIENT TIPS

- Pasteurise: Heat to ~65°C and hold for several minutes. Use a WAPI to check.
- Retained-heat cooking: Boil, then place in insulated container to continue treatment.
- Use solar ovens or reflectors to boil with sunlight.
- Boil larger volumes at once to save fuel.

5. IMPROVING TASTE

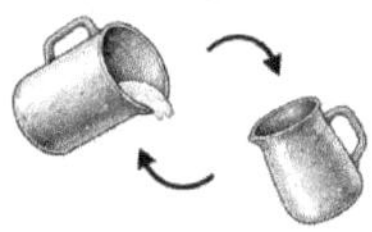

- Boiled water can taste flat or metallic.
- Pour between containers several times to reintroduce air, or add a small pinch of salt.

6. WHAT BOILING DOES NOT DO

- Does not remove heavy metals (lead, arsenic, cadmium, etc.).
- Does not remove agricultural chemicals (pesticides, herbicides).
- Does not remove fuel or hydrocarbons – boiling can concentrate them.
- Does not remove cyanotoxins from algal bloom – heat-stable.

REMEMBER

- Boiling = definitive treatment for biological contaminants.
- Not a solution for chemical contamination.
- If chemical contamination is suspected, use an alternative source, activated carbon filtration, or distillation.

What Boiling Does Not Do

Boiling addresses every biological contaminant category covered in Chapter 15 comprehensively. It does not address:

- **Dissolved heavy metals**, including lead and other contaminants that may be present in water from old plumbing, roof materials, or contaminated groundwater.

- **Agricultural chemicals, pesticides and herbicides**, which remain in the water after boiling and are not broken down by the temperatures involved.

- **Fuel and hydrocarbon contamination**, where boiling can in some cases concentrate certain volatile compounds as the water evaporates, rather than removing them, making boiling actively unsuitable as a response to this specific hazard.

- **Cyanotoxins from blue-green algal bloom**, which are heat-stable and pass through boiling completely unaffected, as set out in Chapters 14 and 15.

Any water source suspected of chemical contamination from these categories requires either an alternative source, or the activated carbon filtration and distillation approaches referenced in Chapter 15, rather than boiling, regardless of how thoroughly it is boiled.

When Boiling Is the Right Choice

Boiling is the appropriate primary treatment method when fuel and a suitable container are available, when the source is assessed as carrying primarily biological rather than chemical risk using the screening approach in Chapter 14, and when time is available to complete the full heating, boiling and cooling cycle. It is less practical for continuous, high-volume treatment of a large household's daily needs over an extended period, where the mechanical filtration methods covered in Chapter 17 generally offer better sustained throughput for the same effort, and is best understood as one tool within the combined, multi-barrier approach that Chapter 19 brings together in full.

Sources referenced in this chapter

National Park Service (US), 2 Ways to Purify Water, nps.gov

Bushwalking Manual Australia, water treatment options, bushwalkingmanual.org.au

World Extreme Medicine, Purifying Wild Water, worldextrememedicine.com

CDC, Making Water Safe in an Emergency, cdc.gov

CHAPTER 17

Mechanical Filtration

Portable mechanical filters represent the current best-practice approach to field water treatment for biological contamination, combining speed, convenience and reliable, repeatable performance without the fuel cost and waiting time that boiling requires. This chapter sets out the specific filter types available, the pore-size ratings that determine what each one actually removes, and the maintenance practices that determine whether a filter continues to perform to its rated standard over its working life.

Pore Size: The Specification That Actually Matters

The single specification that determines a filter's effectiveness is its pore size rating, expressed in microns, and understanding this figure in relation to the contaminant sizes covered in Chapter 15 is the key to choosing correctly.

A filter rated at 0.2 microns reliably removes all bacteria, which range from approximately 0.5 to 10 microns, and all protozoa, including Giardia cysts and Cryptosporidium oocysts, which range from approximately 3 to several hundred microns.

Viruses, ranging from approximately 0.02 to 0.3 microns, are not reliably removed by a 0.2-micron filter, since a meaningful proportion of viral particles are smaller than the filter's pore size and pass through unaffected. This is the single most important limitation of mechanical filtration and the reason it should not be treated as a complete, stand-alone treatment method for any source with a realistic risk of human sewage contamination.

A filter rated finer than 0.2 microns, such as the 0.1-micron hollow fibre membranes used in several premium portable filters, provides a larger safety margin against bacteria and protozoa but still does not reliably address the smaller end of the viral size range.

Virus removal requires either a filter that incorporates an additional chemical or electrostatic disinfection stage, or separate chemical treatment applied after filtration, as covered in Chapter 18 and brought together in the multi-barrier approach in Chapter 19.

Personal Filter Straws

Personal filter straws, including the widely recognized LifeStraw and comparable products from other manufacturers, are designed for direct drinking from a water source, with the filter body itself acting as both the collection point and the filtration device.

Typically rated at 0.2 microns, removing bacteria and protozoa effectively.

Capable of processing hundreds to several thousand litres over their service life, depending on the specific model and the turbidity of the water processed.

Their principal limitation is that they cannot fill a bottle or container for later use or for sharing among a group; they are designed for immediate individual consumption directly from the source.

Best suited to an individual go-bag or bug-out kit, where personal field treatment from whatever source is available, rather than group water provisioning, is the primary use case.

Squeeze and Gravity-Fed Filters

Squeeze filters and gravity-fed filters, produced by manufacturers including Sawyer, Katadyn and MSR, offer considerably better practical flow rates than a personal straw, the ability to fill containers directly, and genuine group water treatment capability.

- Products in this category, such as the Sawyer Squeeze and similar hollow fibre membrane filters, typically filter to approximately 0.1 microns, removing bacteria and protozoa with a meaningful safety margin below the 0.2-micron threshold.

- These filters can be used in several configurations: drinking directly through an attached mouthpiece, squeezing water through into a separate clean container, or suspending the filter as part of a gravity-fed system that processes a larger volume with no ongoing manual effort once set up.

- Typical units weigh 85 to 150 grams and can process up to a million litres before requiring replacement, representing exceptional long-term value relative to their purchase cost.

The critical care requirement: never allow a hollow fibre filter to freeze. Freezing ruptures the microscopic hollow fibres within the membrane in a way that is frequently not visible on inspection but that permanently compromises the filter's ability to exclude bacteria and protozoa to its rated standard. A filter that has been exposed to freezing temperatures, even briefly, should be treated as compromised and either tested for continued integrity according to the manufacturer's guidance or replaced outright, rather than trusted on appearance alone.

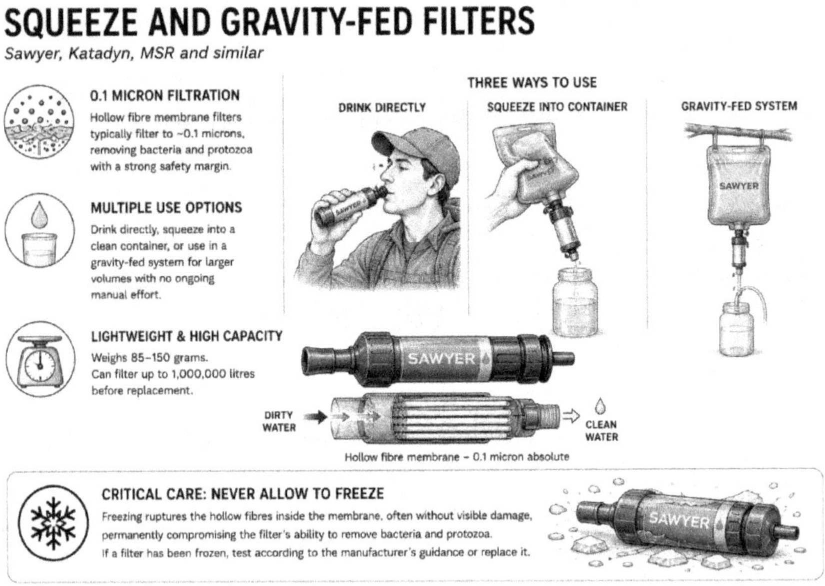

Ceramic Filter Elements

Ceramic filter elements, used in larger portable and base-camp filter units from Katadyn and similar manufacturers, provide filtration performance broadly comparable to hollow fibre membranes, with different practical trade-offs.

- Ceramic elements are considerably more resistant to

chemical degradation and generally have a longer physical service life than hollow fibre membranes.

- They are heavier and bulkier than hollow fibre filters, making them better suited to a base camp, a vehicle kit, or fixed household use than to a lightweight pack-in kit.

- They can be cleaned in the field by gently scrubbing the ceramic surface with a soft brush, which restores flow rate once it has been reduced by particulate build-up, a maintenance advantage over sealed hollow fibre cartridges.

- They are genuinely fragile in a different way to hollow fibre filters: a dropped ceramic element can develop hairline cracks that are not visible on casual inspection but that compromise filtration completely, allowing unfiltered water through the crack.

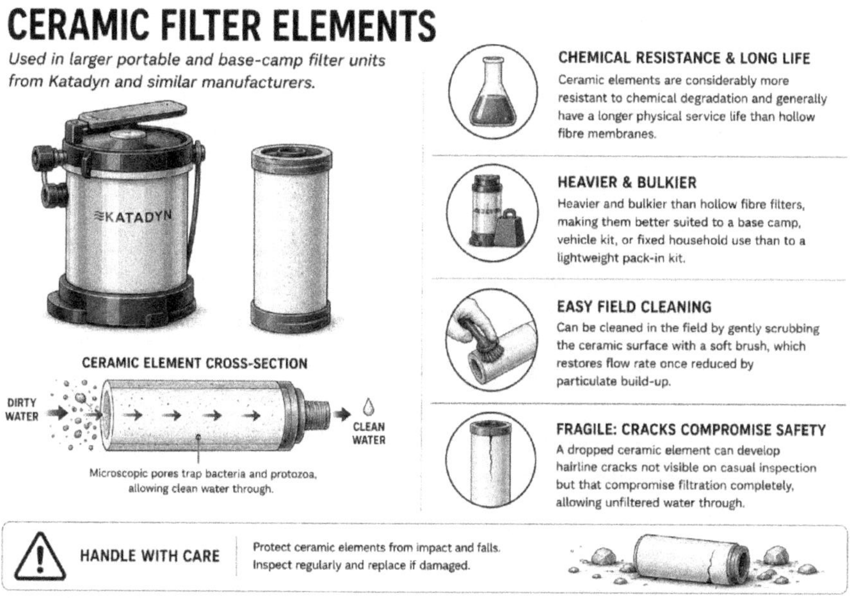

Activated Carbon Stages

Many filter systems, particularly gravity-fed and larger base-camp units, incorporate an activated carbon stage in addition to the mechanical filtration element. Activated carbon does not remove bacteria, protozoa or viruses on its own, and should not be relied upon as a biological treatment stage. Its genuine value lies elsewhere: reducing unpleasant taste and odour, and providing meaningful, though not complete, reduction of some dissolved chemical contaminants, including certain pesticides and, to a lesser extent, some heavy metals. A carbon stage is a useful addition to a mechanical filter system but does not change the underlying virus limitation set out earlier in this chapter, and does not substitute for the dedicated chemical contamination approaches referenced in Chapter 15 where heavy chemical contamination is suspected.

Filter Care and Maintenance

Getting the rated performance out of any filter over its working life depends on consistent maintenance practice, not just the initial purchase decision.

Backflush or clean after each significant use, following the manufacturer's specific method, to restore flow rate as particulate matter accumulates within the filter medium.

Never store a hollow fibre filter wet without either treating it with the manufacturer's preservation solution or allowing it to dry completely first, since standing water left within the filter medium between uses can support bacterial growth inside the filter itself.

Keep hollow fibre filters above freezing at all times, during both storage and use, as set out above.

Track the rated service life or volume for the specific filter and plan replacement before that figure is reached. A filter used well past its rated life may still pass less contamination than an untreated source, but it should not be assumed to provide the full rated level of protection once it has exceeded its specified service life.

Inspect ceramic elements for cracks before each significant trip or period of storage, particularly after any transport where the filter may have been dropped or knocked.

Selecting the Right Filter for the Household

For an individual go-bag or a child's school kit, a personal filter straw provides adequate, low-cost, lightweight protection against the most likely biological hazards. For household or group preparedness, a squeeze or gravity-fed hollow fibre system provides considerably better throughput and versatility for a modest weight and cost increase, and is the more appropriate primary choice for the two-week and extended household planning covered in Part II. For a fixed base camp, a remote property, or any application prioritising long service life and field-repairability over portability, a ceramic system is worth the additional weight and bulk. In every case, filtration should be understood as addressing bacteria and protozoa comprehensively, while requiring a complementary treatment stage for viruses wherever the source carries a realistic risk of human sewage contamination, a combination covered in full in Chapter 19.

Sources referenced in this chapter

Manufacturer specifications for Sawyer, Katadyn, MSR and LifeStraw portable filtration products

Bushwalking Manual Australia, water treatment options, bushwalkingmanual.org.au

National Park Service (US), 2 Ways to Purify Water, nps.gov

CHAPTER 18

Chemical Disinfection

Chemical disinfection is the most compact and weight-efficient water treatment option available, requiring nothing beyond a small quantity of tablets, drops or liquid concentrate. It is the standard treatment method for military field use, expedition use, and as a backup treatment in any kit where weight and space are genuinely constrained. Unlike mechanical filtration, correctly dosed chemical disinfection addresses the full biological spectrum covered in Chapter 15, including viruses, making it the most complete single-method treatment available for field use.

Chlorine Dioxide: The Current Standard

Chlorine dioxide, sold under various product names including specific Aquatabs and Micropur formulations, as tablets, liquid drops, or two-component mixing kits, is effective against bacteria, viruses, Giardia and Cryptosporidium, making it the most complete-spectrum chemical treatment currently available for practical field use.

Standard contact time for clear water above 15 degrees Celsius is typically 30 minutes, though this varies by specific product and concentration, and the manufacturer's instructions for the exact product in use should always take precedence over a general figure.

Cold water requires longer contact time, since the chemical reaction rate driving disinfection is temperature-dependent and slows in cold conditions.

Turbid or coloured water contains organic material that consumes disinfectant before it can act on pathogens, meaning turbid water either needs pre-filtration to remove the organic load, as covered in Chapter 14's screening guidance, or a doubled dosage to compensate.

Cryptosporidium coverage specifically frequently requires an extended contact time, in some product formulations up to four hours or more, considerably longer than the standard 30-minute figure used for bacteria and viruses. Households relying on chemical disinfection as their sole treatment method for a source with a realistic Cryptosporidium risk, such as any source affected by livestock access, should check the specific product's stated Cryptosporidium contact time rather than assuming the standard dosing period is sufficient.

Iodine: A Diminishing Role

Iodine tablets were historically the standard chemical treatment for field water disinfection and remain available, but are now generally less preferred than chlorine dioxide for several specific reasons.

Less complete effectiveness against Cryptosporidium compared to chlorine dioxide at standard field dosing.

A distinctive taste that many people find unpleasant, which matters practically because reduced palatability can lead to reduced fluid intake at exactly the point when maintaining hydration is critical.

Contraindications for pregnant women and for anyone with a thyroid disorder, since iodine is absorbed and metabolised in ways that can affect thyroid function with repeated use.

Potential thyroid effects with extended use even in people without a pre-existing thyroid condition, making iodine considerably less suitable for the extended-duration treatment scenarios covered in Part VIII than for genuinely short-term use.

For a short-term emergency kit where chemical treatment will be used for a matter of days rather than weeks, iodine tablets remain an acceptable backup option. For any extended-use scenario, or for any household including a pregnant member or a member with a thyroid condition, chlorine dioxide is the clearly preferred chemistry.

Household Bleach for Point-of-Use Disinfection

Unscented household chlorine bleach, containing 5 to 9 percent sodium hypochlorite as the only active ingredient, provides an effective, low-cost, widely available alternative to purpose-made tablets for treating water at the point of use, distinct from the long-term storage stabilisation dosing covered in Chapter 5.

- Confirm the bleach is unscented, free of additional cleaning agents or colour-safe formulations, and lists sodium hypochlorite in the 5 to 9 percent range as its only active ingredient.

- Add approximately eight drops of bleach per litre of water, or one-eighth of a teaspoon per gallon, for relatively clear water.

- Stir thoroughly and allow the water to stand for at least 30 minutes before drinking.

- Check for a faint chlorine odour after the standing period. If no odour is detectable, repeat the dosage and allow a further 15 minutes of contact time. If the water still shows no chlorine odour after this second dose, discard it and seek an alternative source, since the absence of any detectable residual suggests the disinfectant has been consumed by a heavier-than-expected organic or chemical load in the water.

- For turbid or visibly contaminated water, pre-filter through the cleanest available cloth first, or double the standard dosage, in the same way as set out for chlorine dioxide products above.

Storage Life of Chemical Treatment Products

Chemical treatment products degrade over time, and confirming a product's condition before relying on it in an emergency is a straightforward but frequently skipped step.

- Most tablet formulations remain effective for four to five years in their original, sealed, unopened packaging when stored in a cool, dry, dark location.
- Once opened, reseal the packaging as completely as practical and use the remaining tablets within the period specified in the manufacturer's instructions, since exposure to air and humidity accelerates degradation once the original seal is broken.
- Tablets stored in humid environments, in a vehicle subject to extreme temperature swings, or loose in an unlabelled bag, may have significantly reduced effectiveness well before their nominal expiry date.
- Always check the expiry date before including a chemical treatment product in any kit, and replace it at expiry regardless of how the tablets appear, since visual inspection cannot confirm remaining chemical potency.

Two-Component Mixing Systems and Silver-Based Products

Beyond single-tablet chlorine dioxide products, two further categories are worth understanding.

Two-component liquid chlorine dioxide kits, requiring the user to mix a measured quantity of two separate liquid reagents before adding them to water, generate chlorine dioxide at the point of use rather than relying on a pre-formed tablet. These systems are widely used in military and expedition contexts because the two-part liquid concentrate has a very long shelf life in its unmixed state and treats a large volume of water per unit of product weight, making it efficient for extended trips or long-term household stockpiling. The trade-off is a slightly more involved preparation step, requiring accurate measurement of both reagents, compared to dropping in a single pre-dosed tablet.

Silver-based products, sold as a water treatment additive in some retail preparedness products, are primarily bacteriostatic, meaning they inhibit bacterial regrowth in already-treated stored water rather than providing the same broad-spectrum, rapid disinfection of contaminated water that chlorine dioxide, iodine or bleach provide. Silver-based products are a reasonable addition for extending the storage life of water already treated by another method, but should not be relied upon as a household's primary disinfection method for treating an untreated or contaminated source.

Choosing Between Chemical Treatment Options: A Summary

For a household's primary chemical disinfection method, chlorine dioxide tablets or an equivalent two-component system provide the most complete protection across bacteria, viruses, Giardia and Cryptosporidium, and are the appropriate default choice for most households.

For a genuinely short-term kit, iodine tablets remain an acceptable and slightly cheaper backup, provided no household member is pregnant or has a thyroid condition.

For point-of-use treatment when no purpose-made product is on hand, unscented household bleach, correctly dosed and given the full contact time, is a fully adequate substitute using a product already in most Australian homes.

For extending the storage life of already-treated water, a silver-based additive is a reasonable supplementary product, but never a substitute for the initial disinfection step.

Combining Chemical Disinfection With Other Methods

Chemical disinfection, correctly dosed for the specific water conditions and contact time required, is capable of serving as a household's sole treatment method where mechanical filtration or a heat source are unavailable. It is nonetheless generally used to best effect as one stage within the combined, multi-barrier approach covered in full in Chapter 19, particularly for turbid natural water sources where pre-filtration substantially improves the reliability of the chemical treatment stage that follows it.

Sources referenced in this chapter

Bushwalking Manual Australia, water treatment options, bushwalkingmanual.org.au

CDC, Making Water Safe in an Emergency, cdc.gov

Ready.gov, Water, ready.gov/water

Manufacturer specifications for Aquatabs and Micropur chlorine dioxide products

CHAPTER 19

UV Treatment and Multi-Barrier Systems

This chapter closes Part VI by covering ultraviolet disinfection, briefly addressing reverse osmosis and desalination, and then bringing every treatment method covered in Chapters 16 through 18 together into a single, coherent framework for treating water sources of genuinely uncertain or elevated risk, which describes the majority of natural water a household will actually encounter during a real disruption.

Ultraviolet Treatment

Ultraviolet light disrupts the DNA of biological organisms, preventing them from reproducing and effectively neutralizing bacteria, viruses and protozoa when the light reaches them directly at a sufficient dose. Portable UV treatment devices, most commonly small pen-sized units such as the SteriPen and similar products, have become a genuinely practical field option over the past two decades.

- A typical UV treatment pen treats a litre of clear water in approximately 90 seconds and can process hundreds of litres on a single battery charge.
- UV treatment is effective against the full biological spectrum covered in Chapter 15, including viruses, addressing the specific gap left open by mechanical filtration alone.
- **The critical limitation is turbidity.** UV light must reach an organism directly to disrupt it, and in turbid or coloured water, suspended particles and dissolved organic material block the light before it can penetrate throughout the full volume being treated. Any water with visible cloudiness or

colour must be clarified, either by settling or by pre-filtration through cloth or a mechanical filter, before UV treatment is applied, or the treatment cannot be relied upon to reach the entire volume.

- UV treatment requires battery power, which makes it less resilient than chemical or mechanical treatment options for a genuinely extended field scenario where recharging is not possible, and battery failure removes the household's access to this treatment method entirely at the point it is most needed.

- For a household emergency kit, a UV pen is a strong addition alongside, rather than a replacement for, the boiling, filtration and chemical methods already covered, since it provides very rapid treatment of a relatively clear source when a trusted water supply is available but the specific tap or tank may be compromised, such as during a boil-water notice.

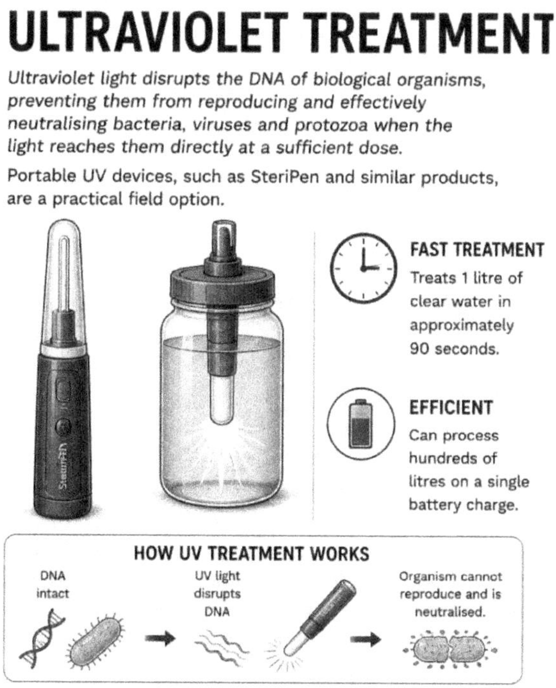

Reverse Osmosis and Distillation

For situations involving genuinely severe chemical contamination, high salinity, or brackish groundwater, mechanical filtration, chemical disinfection and UV treatment, covered so far in Part VI, all share a common limitation: none of them remove dissolved salts or most dissolved chemical contaminants.

Reverse osmosis forces water through a semi-permeable membrane under pressure, removing dissolved salts, most heavy metals, and a wide range of chemical contaminants, alongside all biological contamination. It is the technology behind both large-scale municipal desalination plants and smaller, portable manual or powered units used for marine survival, covered further in Part VIII.

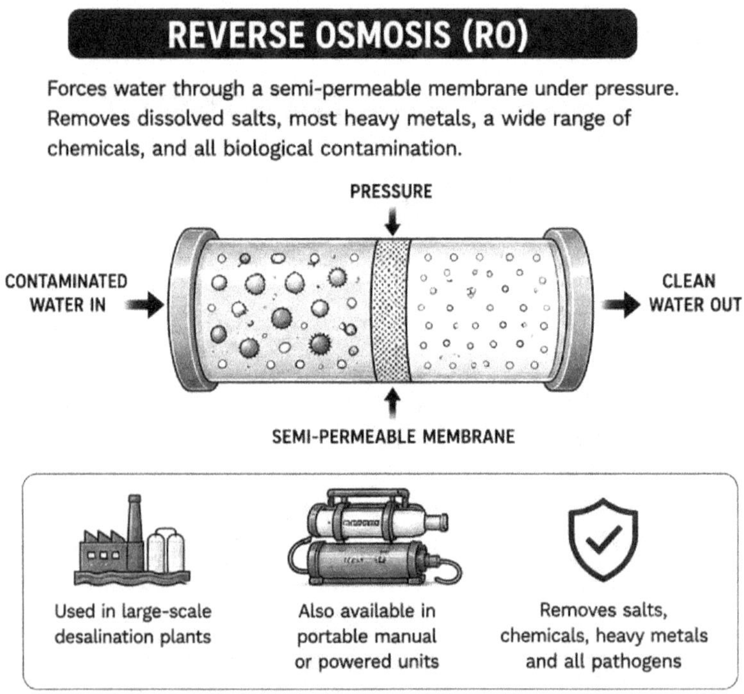

Distillation, as introduced through the solar still in Chapter 9, achieves a similar practical result by evaporating water and condensing the vapour separately from the source, leaving salts, heavy metals and non-volatile organic contamination behind.

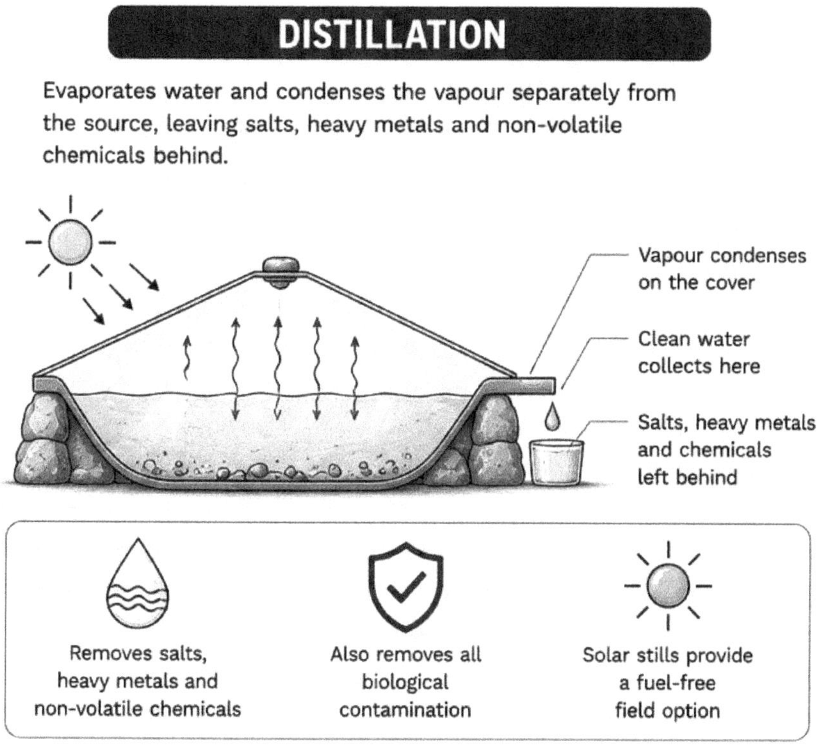

Both approaches are considerably more equipment-intensive, slower, or more expensive than the methods covered in Chapters 16 through 18, and are generally reserved for situations where the water source is known or suspected to carry chemical or salinity contamination that the standard biological treatment methods cannot address, rather than being used as a routine household approach.

Building the Multi-Barrier Approach

For any water source assessed, using the screening framework in Chapter 14, as carrying an elevated or uncertain risk, whether from turbidity, proximity to human or livestock activity, or an unknown contamination history, the most reliable treatment approach combines multiple methods in sequence, each addressing a different category of risk that a single method alone would miss.

Coarse pre-filtration. Pass the water through the cleanest available cloth, a bandana, a t-shirt, or a dedicated pre-filter, to remove visible sediment and organic material. This step protects the effectiveness of every subsequent stage: sediment shields pathogens from chemical and UV treatment, and quickly clogs fine mechanical filters.

Mechanical filtration. Pass the pre-filtered water through a portable filter rated at 0.2 microns or finer, removing bacteria and protozoa comprehensively, as covered in Chapter 17.

Chemical disinfection. Add chlorine dioxide, or an equivalent chemical treatment, at the correct dosage and full contact time for the water's temperature and clarity, to address viruses and any bacteria that may have passed through a compromised or ageing filter, as covered in Chapter 18.

Optional UV treatment, once the water is already clarified through the earlier stages, can be added as a further, very rapid confirmation step, particularly useful where a fourth layer of assurance is wanted for a vulnerable household member covered in Chapter 20.

The outcome of this combined sequence is water that has had every biological threat category covered in Chapter 15 addressed through at least one, and in most cases two, independent treatment stages. This layered redundancy is precisely what a single treatment method, however well executed, cannot provide on its own.

Matching the Combination to the Source

Not every source requires the full four-stage sequence, and applying unnecessary treatment steps to a genuinely low-risk source wastes time and consumable treatment products without meaningfully improving safety.

A relatively clean, flowing highland stream, with no visible upstream agricultural activity, livestock access or human settlement, and assessed as low risk under the Chapter 14 screening framework, can generally be treated adequately with mechanical filtration alone, or boiling alone, without the full combined sequence.

Agricultural runoff-affected lowland streams, still water in flood conditions, water from a hand-dug emergency well, or any source in reasonable proximity to human settlement or livestock, should receive the full combined approach: pre-filtration, mechanical filtration and chemical disinfection at minimum.

Any source where blue-green algal bloom is suspected, as covered in Chapters 14 and 15, should not be treated at all using any method in this book. The multi-barrier approach addresses biological risk comprehensively; it does not address cyanotoxins, and no amount of additional treatment stages changes this.

When in doubt about which category a source falls into, apply the more thorough combination rather than the lighter one. The additional cost is a few minutes of time and a small quantity of consumable treatment product. The cost of under-treating is a waterborne illness at precisely the point in a disruption when the household can least afford to manage one.

Part VI in Summary

Boiling, mechanical filtration, chemical disinfection and UV treatment each address the biological contaminants covered in Chapter 15 through a different physical or chemical mechanism, and each has a specific, well-defined limitation: boiling and chemical disinfection require time and, in the case of boiling, fuel; mechanical filtration alone leaves viruses largely untreated; UV treatment requires clarity and battery power; and none of these four methods addresses dissolved chemical contamination or cyanotoxins. Understanding these mechanisms and limitations, rather than memorising a single preferred method, is what allows a household to make a genuinely informed treatment decision for whatever specific source it is actually facing, rather than applying the same routine regardless of the risk profile involved.

Sources referenced in this chapter

Manufacturer specifications for SteriPen and similar portable UV treatment devices

Bushwalking Manual Australia, water treatment options, bushwalkingmanual.org.au

National Park Service (US), 2 Ways to Purify Water, nps.gov

General technical references on reverse osmosis and desalination for water treatment

CHAPTER 20

Vulnerable Household Members

Every calculation, treatment method and storage system covered so far in this book has been framed around the needs of a generally healthy adult. Several categories of household member face meaningfully different risks from both water scarcity and water contamination, and a genuinely complete household water plan addresses these differences explicitly rather than assuming a single standard applies equally to everyone in the home.

Infants and Young Children

Infants and young children face a combination of higher fluid requirements relative to body weight, less mature immune systems, and, for formula-fed infants, a complete dependence on correctly prepared water for their entire nutritional intake.

Formula preparation. Water used to prepare infant formula should always be treated to the highest standard available in the household, using boiling as the preferred method given its complete reliability against every biological contaminant covered in Chapter 15, regardless of what level of treatment is considered adequate for the water used by adult household members from the same source.

Cooling before use. Boiled water intended for formula preparation should be cooled to an appropriate temperature before mixing, following standard infant feeding safety guidance, and stored appropriately if not used immediately.

Nitrate risk. As covered in Chapter 15, any bore or well water used for infant formula, particularly in an agricultural area, carries a specific risk from nitrate contamination that can impair an infant's oxygen-carrying capacity. Where a household relies on bore water and has an infant in the home, testing that water specifically for nitrate levels, in addition to the general biological testing covered in Chapter 14, is a reasonable and worthwhile precaution.

Higher relative fluid requirement. As set out in Chapter 2, children require less absolute volume than adults but proportionally more relative to body weight, and are less reliable than adults at recognising and reporting their own thirst, meaning a caregiver should actively monitor and prompt fluid intake during any period of heat or disruption rather than relying on the child to ask.

Pregnant and Breastfeeding Women

Pregnancy and breastfeeding both increase fluid requirements measurably above the standard adult baseline set out in Chapter 2, and introduce a specific complication for chemical water treatment.

Increased fluid requirement, as set out in Chapter 2, of approximately an additional litre per day above standard adult intake for both pregnancy and breastfeeding, reflecting the genuine physiological demand of supporting a pregnancy or milk production.

Iodine tablets should be avoided as a chemical treatment method during pregnancy, as covered in Chapter 18, given the documented effect of iodine on thyroid function. Chlorine dioxide, boiling, or mechanical filtration are all appropriate alternative treatment methods that do not carry this specific concern.

Heat sensitivity. Pregnant women are at increased risk of heat-related illness compared to the general adult population, reinforcing the importance of the heat adjustment to water planning figures set out in Chapter 2 for any household including a pregnant member during Australian summer conditions.

Elderly Household Members

Older adults face a specific and frequently underappreciated risk: a blunted thirst response that can allow meaningful dehydration to develop well before the individual feels thirsty enough to seek water on their own initiative.

Do not rely on thirst as the indicator of an elderly household member's hydration status. Instead, establish a specific, planned fluid intake schedule and actively monitor whether it is being met, particularly during a heatwave or any period of disruption when normal routines and reminders may be interrupted.

Medication interactions. Some medications commonly taken by older adults, including certain diuretics and blood pressure medications, affect fluid balance directly, and a household with an elderly member taking such medication should factor this into daily water planning rather than treating every household member's requirement as identical.

Mobility considerations affect access to water during a disruption in a way that is easy to overlook: an elderly household member with reduced mobility may be less able to reach a distribution point, carry a full jerry can, or access water stored in an awkward location, and the storage location guidance in Chapters 3 and 4 should account for this directly, storing at least some portion of the household reserve somewhere genuinely accessible to every household member regardless of mobility.

Immunocompromised Household Members

Anyone who is immunocompromised, including people undergoing chemotherapy, people who are HIV-positive with a low CD4 count, and organ transplant recipients, faces a meaningfully higher risk from water-borne pathogens that a healthy adult would tolerate without becoming seriously ill.

Lower the treatment threshold, not just the treatment standard. For an immunocompromised household member, water should be treated as a matter of routine in any circumstance where source integrity is uncertain, rather than only during a formally declared boil-water advisory or an obviously compromised source. This applies to the rainwater tank guidance in Chapter 6 as much as to any natural water source covered in Part III.

The consequences of exposure are more serious, not merely more likely, for this group, meaning the multi-barrier approach set out in Chapter 19 is the appropriate default treatment standard for this household member specifically, even where a lighter single-method treatment might be considered adequate for other, healthy household members drinking from the same source.

Coordinate with the treating medical team. Anyone managing an immunocompromising condition or treatment should discuss water safety specifically with their treating clinician as part of broader emergency planning, since individual risk varies considerably depending on the specific condition and its current management.

Dialysis Patients

Dialysis represents a distinct category of water-related medical vulnerability that sits outside the scope of household-level treatment entirely.

Dialysis requires water purified to a considerably higher standard than any method covered in Part VI of this book can achieve. Standard boiling, filtration and chemical disinfection address biological contamination but do not remove the trace chemical contaminants that dialysis-grade water purification is specifically designed to exclude, since dialysis involves direct exposure of the bloodstream to the water used in the process.

This is a clinical planning problem, not a household preparedness problem. Dialysis patients and their families should develop a specific emergency plan directly with their treating dialysis clinic or hospital, addressing continued clinic access during a disruption, any home dialysis options and their specific water requirements, and clear communication protocols for an extended disruption. No guidance in this book should be understood as a substitute for that clinical planning relationship.

Other Medically Complex Household Members

Several further conditions warrant specific consideration within a household's water plan.

Diabetes can affect fluid balance and increase susceptibility to the consequences of dehydration, and a household including a person managing diabetes should apply the heat and activity adjustments from Chapter 2 attentively rather than treating this household member as a standard case.

Kidney disease, short of dialysis dependency, can affect both fluid requirements and the ability to safely process certain minerals present in some treated or bore water, and any household including a member with significant kidney impairment should seek specific guidance from their treating clinician on both fluid volume and water quality requirements.

Any household member on regular medication requiring water for administration should have that specific, non-negotiable requirement accounted for explicitly in the medication component of the daily water calculation set out in Chapter 2, rather than assumed to be covered by the general household drinking water allowance.

Pets, Working Animals and Livestock

As set out in Chapter 2, animals in the household carry their own distinct water requirements that should never be assumed to be covered incidentally by the human household allowance. A working dog, an elderly or unwell pet, or livestock kept for food production each carry a specific vulnerability profile in the same way human household members do, and the same principle of not relying on the animal's own behaviour as a reliable indicator of need applies here as it does for elderly and very young human household members: monitor and plan proactively rather than waiting for visible signs of distress.

Bringing Vulnerability Planning Into the Household Plan

None of the guidance in this chapter requires a separate water system for each vulnerable household member. It requires the household's existing storage, sourcing and treatment plan, covered across Parts II through VI, to be applied with the specific adjustments set out above: higher treatment standards, more conservative thresholds for when treatment is applied at all, and explicit rather than assumed accounting for the additional volume and quality requirements these household members carry. Chapter 23 brings this together into a single, written household water security plan that makes these adjustments explicit and repeatable rather than something to be worked out under pressure during an actual disruption.

Sources referenced in this chapter

Australian Drinking Water Guidelines (NHMRC), vulnerable population considerations

CDC and Ready.gov, guidance on water needs for infants, pregnant women and immunocompromised individuals

CHAPTER 21

Treating Water-Related Illness

Every method in this book is designed to prevent water-related illness before it starts. Despite the best preparation, illness from contaminated water, or from inadequate water intake, can still occur, whether from a source that was misjudged, a treatment step that was skipped under pressure, or simple bad luck. This chapter sets out how to recognise the two broad categories of water-related illness, dehydration and waterborne infection, and how to respond appropriately, including the point at which home management is no longer sufficient and professional medical care is required.

This chapter provides general information for planning purposes and is not a substitute for professional medical advice. Where any doubt exists about the severity of an illness, particularly in an infant, an elderly person, a pregnant woman, or anyone immunocompromised, seek medical care rather than relying on home treatment alone.

Recognising Dehydration

Chapter 1 set out the physiological progression of dehydration as a percentage of body weight lost as fluid. In a practical setting, without a scale or a clinical assessment available, the following signs provide a usable field guide to severity.

Mild dehydration. Thirst, a dry mouth, and dark yellow urine. This stage is straightforward to correct with oral fluid intake alone.

Moderate dehydration. Reduced urine output, noticeable fatigue, headache, dizziness on standing, and a further darkening of urine colour toward amber. Active rehydration, ideally with an oral rehydration solution as set out below rather than plain water alone, should begin immediately.

Severe dehydration. Confusion or irritability, sunken eyes, minimal or absent urine output, rapid heart rate, and, in infants and young children, a sunken fontanelle (the soft spot on a baby's head) and a notable lack of tears when crying. Severe dehydration is a medical emergency requiring urgent professional care, and rehydration efforts should not delay seeking that care.

Oral Rehydration Solution

Plain water alone does not always adequately replace the dehydration associated with vomiting or diarrhoea, since these conditions cause the body to lose salts and sugars alongside fluid, and replacing fluid without also replacing these components can be less effective than a properly balanced solution.

Commercial oral rehydration products, available from pharmacies, should be the first choice where available and are formulated to the correct balance for effective rehydration. Every household preparedness kit benefits from including a supply of these products alongside its water storage and treatment equipment.

A basic home-made oral rehydration solution, for use when commercial products are unavailable, can be prepared using the following widely used formula, based on World Health Organization guidance for oral rehydration:

Start with one litre of clean, treated water, prepared using the methods in Part VI.

Add six level teaspoons of sugar.

Add half a level teaspoon of salt.

Stir thoroughly until fully dissolved.

Sip slowly and steadily rather than drinking a large volume at once, particularly where nausea is present, since a large volume taken quickly can trigger vomiting and worsen the fluid loss the solution is intended to correct.

This solution should be freshly prepared and not stored for an extended period once made, and should be discarded and a fresh batch prepared if not used within 24 hours.

Signs of Specific Waterborne Illnesses

Recognising the likely cause of an illness helps inform the appropriate response, though a definitive diagnosis generally requires medical assessment rather than home judgement alone.

Giardiasis, caused by the Giardia parasite covered in Chapter 15, typically presents with diarrhoea, abdominal cramping, bloating, gas and fatigue, often beginning one to three weeks after exposure and persisting for several weeks if untreated. It rarely causes fever.

Cryptosporidiosis, caused by Cryptosporidium, presents similarly to giardiasis with watery diarrhoea, cramping and, in some cases, a low-grade fever, typically beginning within two weeks of exposure.

Bacterial gastroenteritis, from organisms including Salmonella and Campylobacter covered in Chapter 15, tends to present more acutely, with a shorter incubation period, often including fever, and sometimes blood in the stool in more severe cases.

Viral gastroenteritis, from norovirus, rotavirus or hepatitis A, presents with a range of severity from mild, short-lived symptoms to, in the case of hepatitis A specifically, more prolonged illness including jaundice in more severe cases.

General Home Management

For a person with suspected water-related gastrointestinal illness who is not showing signs of severe dehydration or other warning signs covered below, the following general measures support recovery.

- Begin oral rehydration immediately, using a commercial product or the home-made solution above, in small, frequent sips rather than large volumes.

- Continue to offer normal food once appetite returns, rather than withholding all food for an extended period, which is no longer standard guidance for most cases of gastrointestinal illness.

- Rest, and avoid unnecessary physical exertion, particularly in hot conditions, until symptoms have substantially resolved.

- Maintain scrupulous hand hygiene for the ill person and everyone else in the household, since several of the organisms covered above spread readily between household members in close quarters, particularly where water for handwashing is itself limited.

- Isolate, where practical, any water source suspected of causing the illness from further household use until it has been properly retested or an alternative source secured, to prevent repeated exposure of the same or other household members.

When to Seek Medical Care

Home management is appropriate for mild to moderate illness in an otherwise healthy adult. The following signs indicate that professional medical care should be sought rather than relying on home treatment alone.

- Signs of severe dehydration as set out earlier in this chapter, in any household member.
- Any symptoms in an infant, particularly reduced feeding, lethargy, a sunken fontanelle, or a significant reduction in wet nappies.
- Blood visible in vomit or stool.
- A high or persistent fever that does not respond to appropriate care.
- Symptoms persisting beyond several days without improvement, or worsening despite home rehydration efforts.
- Any symptoms in a pregnant woman, an elderly person, or anyone immunocompromised, dialysis-dependent, or otherwise medically vulnerable as covered in Chapter 20, given the generally higher risk of complications in these groups.
- Signs of significant abdominal pain beyond ordinary cramping, which can indicate a more serious underlying condition unrelated to simple gastrointestinal infection.

Preventing Recurrence

Once an illness has occurred, or been narrowly avoided, treat it as a signal to review the household's water sourcing and treatment practices rather than an unlucky, isolated event. Revisit the screening framework in Chapter 14 for the specific source involved, confirm the treatment method used was appropriate for that source's assessed risk level using the guidance in Chapter 19, and check that any equipment involved, particularly a mechanical filter or a chemical treatment product, was within its rated service life and had not been compromised, such as by freezing, as covered in Chapter 17. A single illness event is frequently the clearest, if least welcome, feedback a household receives about a gap in its water safety practices, and is worth acting on directly rather than dismissing as bad luck alone.

Sources referenced in this chapter

World Health Organization, oral rehydration solution formula and guidance

Australian Drinking Water Guidelines (NHMRC), waterborne pathogen clinical profiles

General clinical guidance on recognising and responding to dehydration in infants and adults

This chapter covers a sensitive health topic. If you or someone in your household is currently unwell, please seek advice from a doctor, pharmacist, or, in a medical emergency, contact emergency services directly.

CHAPTER 22

Rationing and Conservation Under Scarcity

When a household's water supply is genuinely limited, the order in which that supply is used matters as much as the total volume available. A household that distributes its remaining water without a clear sense of priority risks using it on lower-value purposes early, only to find the higher-value uses, drinking water and medical needs, under-supplied later when the disruption has not yet resolved. This chapter sets out how to prioritise water use deliberately, how to track consumption against remaining supply, and the specific point at which rationing drinking water itself becomes dangerous rather than prudent.

The Priority Order for Limited Water

Rather than a list of ranked rules recited from first to last, the priorities below are better understood as three tiers, each with a clear rationale, so a household under pressure can make a fast, sound decision about where its next litre of water should go.

Tier one: non-negotiable, life-sustaining uses.

- Drinking water, at the minimum rate required to prevent heat illness and maintain cognitive function, generally at least two litres per person per day in moderate conditions and considerably more in heat, as set out in Chapter 2.
- Medication preparation for any household member taking medication that requires water for safe administration.
- Wound cleaning, since inadequate wound care in a disrupted environment with limited subsequent medical access is a direct and serious infection risk.

- Infant formula preparation, where quality and hygiene standards cannot be compromised regardless of overall supply pressure.

Tier two: important uses that reduce illness risk and maintain basic function.

- Handwashing, specifically before food handling and after using the toilet, since this single practice does more to prevent the spread of illness through a stressed household than almost any other water use.

- Basic food preparation, including rehydrating stored food and cooking staples.

- Minimal personal hygiene beyond handwashing, using the conservation techniques covered in Chapter 12, such as sponge bathing rather than a full shower.

Tier three: lower-priority uses to reduce or eliminate first when supply tightens.

- Toilet flushing, which can generally be shifted to greywater or bathtub pre-fill water, as covered in Chapters 3 and 12, rather than drawn from the drinking water reserve.

- General cleaning, beyond the minimum needed to prevent illness.

- Any discretionary use, including watering ornamental plants, washing vehicles, or hosing hard surfaces, which should be eliminated entirely once supply is under genuine pressure.

The practical rule this ordering produces is straightforward: when supply tightens, cut from tier three first, then tier two, and treat tier one as fixed and non-negotiable regardless of how limited the remaining supply becomes.

The One Rule That Should Never Be Broken

Every reputable emergency planning body, including agencies referenced throughout this book, is consistent on a single point: never ration drinking water below the minimum required to prevent heat illness and maintain basic function. Where a genuine shortfall exists, the correct response is to cut every lower-priority use in tiers two and three completely, and to actively pursue the additional sourcing methods covered in Parts III and IV, rather than to stretch the drinking water allowance below a safe minimum. Rationing drinking water itself is not a form of prudent conservation; it is the specific failure mode this entire book is designed to prevent, and it should be understood as a last resort only once every other reduction has already been made and additional sourcing has genuinely been exhausted or is not yet available.

Tracking Consumption Actively

A household facing a genuine water shortage benefits enormously from tracking its position explicitly rather than estimating informally, since the difference between an accurate and an inaccurate sense of remaining supply can be the difference between a manageable situation and a genuine crisis.

Know the starting volume. Record the total litres in storage at the point a disruption begins, using the containers and volumes established in Part II.

Track daily usage. Even an approximate daily log, noting roughly how many litres were used and for what purpose, provides the information needed for every subsequent decision in this section.

Calculate remaining days at the current rate. Divide the remaining stored volume by the current daily usage rate to produce a straightforward estimate of how many days the supply will last if nothing changes.

Compare that figure against the likely duration of the disruption, using whatever official information is available, and adjust usage immediately if the current rate would exhaust supply before the disruption is expected to resolve.

As a concrete illustration: a household that begins a disruption with 100 litres in storage and uses 30 litres per day will exhaust that supply in three to four days. The same household, applying the tier-based conservation approach set out in this chapter to bring daily use down to 12 litres per day, extends the same 100-litre supply to more than eight days. The difference between these two outcomes is entirely within the household's control, and is determined by whether consumption is being tracked and managed deliberately from the first day of the disruption, rather than left to informal habit.

Deciding When to Activate Additional Sourcing

Water tracking is only useful if it feeds into a clear decision about when to move beyond the household's stored reserve and begin actively pursuing the additional sources covered in Parts III and IV.

A reasonable trigger point is when the calculated remaining-days figure falls below the expected remaining duration of the disruption, based on the best available official information, even after every tier-two and tier-three reduction in this chapter has already been applied.

Acting early is considerably safer than acting late. Beginning to harvest rainwater, capture greywater, or assess natural sources while the household still has a meaningful drinking water buffer allows time to screen and treat any additional source properly, using the methods in Parts V and VI, rather than being forced into a rushed, higher-risk decision once the stored reserve is nearly exhausted.

Do not wait for the stored reserve to reach zero before beginning this process. The households that manage an extended disruption most successfully are consistently the ones that treat the transition to additional sourcing as a planned, proactive step taken with a buffer still in hand, not a reactive scramble triggered by the last container running dry.

Communicating the Plan to the Household

A rationing and conservation plan is only effective if every household member understands and follows it, and this requires clear, simple communication rather than an assumption that everyone will independently arrive at the same sensible conclusions under pressure.

Explain the tier system in plain terms to every adult and older child in the household, so that decisions about water use during a genuine shortage are consistent across everyone drawing from the same limited supply.

Post a simple, visible reminder near the household's main water storage location during an active disruption, covering the current daily allowance and the priority order, so the plan does not rely purely on memory during a stressful period.

Revisit and adjust the plan as circumstances change, including as the disruption's likely duration becomes clearer, as additional sourcing comes online, or as any household member's needs change, such as illness or the arrival of a house guest during the disruption.

Sources referenced in this chapter

Ready.gov, Water, ready.gov/water: guidance on never rationing drinking water unless directed by authorities

CDC, How to Create and Store an Emergency Water Supply, cdc.gov/water-emergency

CHAPTER 23

Building Your Complete Water Security Plan

This chapter brings together the calculations, systems and decision frameworks covered across the preceding twenty-two chapters into a single, written household water security plan. A plan that exists only as scattered knowledge, however thorough, is considerably less useful under pressure than the same knowledge captured in a document the household can refer to directly. This chapter sets out how to build that document, staged appropriately to the household's actual risk profile and budget, and how to keep it current over time.

Step One: Calculate the Household's Real Requirement

Begin with the method set out in Chapter 2. List every household member, every pet and animal, and calculate a daily water requirement in normal conditions, then apply the appropriate heat and activity adjustments. This single figure, expressed in litres per day, is the foundation every later step in this chapter multiplies out to produce specific storage and sourcing targets.

Step Two: Choose a Preparedness Tier

Using the mild, moderate and extreme framework introduced in Chapter 1, decide realistically which tier the household is planning for, recognising that this decision should be revisited periodically rather than fixed permanently.

Mild tier: a properly stocked 72-hour kit, as covered in Chapter 3, sized to the household's calculated daily requirement multiplied by three, plus at least one purpose-built bathtub bladder for any household in a flood-prone or supply-vulnerable area.

Moderate tier: a full two-week reserve, as covered in Chapter 4, supplemented by a rainwater harvesting system sized to genuine emergency-reserve capacity as set out in Chapter 6, and a basic household greywater capture practice as covered in Chapter 12.

Extreme tier: the moderate tier in full, plus long-term bulk storage as covered in Chapter 5, a licensed bore where the property and budget support it as covered in Chapter 7, a complete field treatment kit spanning boiling, filtration and chemical disinfection as covered in Chapters 16 through 19, and familiarity with the natural sourcing techniques in Part III and the extreme scenario material in Part VIII.

A household does not need to commit to the extreme tier to be genuinely well prepared. Solid execution of the mild and moderate tiers covers the overwhelming majority of realistic Australian household disruptions, and this should be the honest starting recommendation for most readers rather than an aspiration toward maximum preparedness for its own sake.

Step Three: Build in Budget Stages

Water security does not need to be built in a single expensive purchase, and staging the investment over time is both financially realistic and, in practice, how most households actually build genuine preparedness.

Immediate, near-zero-cost actions: calculate the household figure from Step One, identify and clean suitable containers already in the home, and establish the household's bathtub pre-fill procedure and hidden-source inventory from Chapter 11.

Low-cost, near-term purchases: food-grade jerry cans and commercial bottled water to complete the 72-hour kit, a basic chemical disinfection product, and a personal filter straw for each household member's go-bag.

Moderate investment: one or two large food-grade storage drums to complete the two-week reserve, a quality squeeze or gravity-fed filtration system for the household, and a purpose-built bathtub bladder or two.

Larger investment, staged over one to several years: a rainwater tank sized to genuine emergency-reserve or primary-supply capacity, an installed greywater diversion system, and, where the property supports it, a licensed bore.

Step Four: Document the Household's Specific Sourcing Priority

Using the material in Parts III and IV, write down the specific sources available to the household, in the order they should be used once stored supply runs low, tailored to the actual property and its surroundings rather than the generic order presented in this book.

- Note the location and estimated yield of any rainwater tank, and the location of the mains shut-off valve.
- Note whether a bore or well exists on the property, and where the nearest known natural water source, such as a creek or dam, sits relative to the home.
- Note which hidden household sources from Chapter 11, such as the hot water system and any pool, are present and what volume each represents.
- Note which greywater capture practices from Chapter 12 the household already uses, or intends to activate, during a disruption.

Step Five: Confirm the Treatment Kit Matches the Sourcing Plan

A sourcing plan is only as good as the household's ability to treat what it finds. Cross-check the treatment equipment on hand, boiling capability, a mechanical filter, a chemical disinfection product, and, where budget allows, a UV pen, against the specific sources identified in Step Four, and address any gap directly rather than discovering it during an actual disruption. A household planning to rely on a bore or rainwater tank as a secondary source, for example, should already have the field test strips from Chapter 14 and the full multi-barrier capability from Chapter 19 on hand, not merely a single treatment method assumed to be sufficient.

Step Six: Note Every Vulnerable Household Member's Specific Adjustment

Working from Chapter 20, write down explicitly, next to each vulnerable household member's name, the specific adjustment their needs require: a higher treatment threshold, an increased daily volume, a specific medical coordination step such as a dialysis clinic's emergency contact details, or a note that iodine tablets must not be used for a pregnant household member. This step converts general guidance into a specific, actionable reference rather than something to be remembered accurately under pressure.

Step Seven: Write the Household's Rationing Rule

Using Chapter 22's tier framework, write a short, specific version of the priority order for the household's own reference, including the calculated daily minimum drinking water figure below which rationing should never go, and post it near the household's main storage location for use during an actual disruption.

Regional Considerations Within the Plan

Australia's geographic and climatic diversity means the same general plan should be weighted differently depending on where a household is located.

Tropical northern Australia, including Far North Queensland, should weight the plan toward cyclone and flood preparedness: robust rainwater tank first-flush and mosquito-proofing given the region's tropical disease profile, and awareness of the coconut water source covered in Chapter 10 as a genuinely relevant local option.

The arid interior should weight the plan toward the extended-duration sourcing and extraction techniques in Part III, given the greater distance to resupply and the more severe consequences of underestimating storage needs during drought, as covered in Chapter 6's rainwater sizing guidance for rural properties.

Temperate southern and coastal Australia should weight the plan toward heatwave-adjusted daily requirements as covered in Chapter 2, and toward bushfire-triggered rainwater tank inspection as covered in Chapter 6.

Flood-prone regions of Queensland and New South Wales should weight the plan toward the bathtub pre-fill timing discipline covered in Chapter 3 and the boil-water notice discipline covered in Chapter 14, given the specific, well-documented pattern of contamination rather than absence following flood events in these regions.

The Annual Review

A written plan that is never revisited gradually becomes inaccurate as household composition, equipment and circumstances change. Set a fixed, memorable annual date, such as the start of the Australian storm season, to review the entire plan.

- Recalculate the household figure from Step One if household composition, including pets, has changed.
- Physically inspect stored water containers, drums and any bathtub bladders, following the inspection guidance in Chapters 3 through 5.
- Check the expiry and service status of every chemical treatment product and mechanical filter, following the guidance in Chapters 17 and 18.
- Confirm the rainwater tank and any greywater system have received their scheduled maintenance from Chapters 6 and 12.
- Update the vulnerable household member notes from Step Six to reflect any change in medical circumstances, medication or household composition.
- Revisit the preparedness tier chosen in Step Two and confirm it still reflects the household's actual risk profile and budget.

The Plan in One Sentence

Every element of this book, storage, harvesting, sourcing, greywater, screening, treatment and the specific needs of vulnerable household members, exists to answer one question honestly at any given moment: how many days of clean water does this household actually have access to right now, and what is the plan to extend that number if it needs extending. A household that can answer that question specifically, rather than approximately, has built genuine water security, regardless of which preparedness tier or budget stage it has reached.

Sources referenced in this chapter

Synthesis of Australian Government, state health department and CDC guidance referenced throughout Chapters 1 through 22 of this book

CHAPTER 24

Stranded and Isolated

Part VIII covers water strategy for circumstances well outside the everyday household disruptions covered in Parts I through VII: genuine isolation, with no stored reserve, no household infrastructure, and often no certainty about when help will arrive. This chapter addresses isolation on land and at sea, whether from a vehicle breakdown in remote Australia, a shipwreck, a disabled vessel adrift, a lost multi-day hike, a plane crash in remote terrain, or deliberate movement through wilderness while avoiding contact with others. Every technique referenced here draws on the detailed methods already covered in Parts III, V and VI; this chapter's contribution is showing how those methods combine under the specific constraints each of these situations imposes.

Vehicle Breakdown in Remote Australia

A vehicle breakdown far from towns, mobile coverage and passing traffic is one of the most common ways Australians actually die from dehydration, and the pattern behind the majority of these deaths is well documented and entirely avoidable.

Stay with the vehicle. This is the single most important rule in outback survival, repeated consistently across every credible source on the subject. A vehicle is dramatically easier for search crews to spot from the air or ground than a person on foot, it provides shelter and shade, and walking away consumes exactly the fluid reserve a stranded person cannot afford to lose. Leaving a vehicle to seek help on foot is strongly associated with fatal outcomes; staying with it is strongly associated with survival.

Reduce fluid loss actively rather than relying on the water alone. Deploy an awning, tarp or reflective space blanket for shade, dig a shallow trench beneath the vehicle for a cooler resting space if the ground allows it, and avoid all unnecessary movement and exertion during the hottest part of the day, generally 11am to 4pm.

Source water from the vehicle itself, using the guidance in Chapter 11. A running air conditioning system produces condensate that is safe to drink and can be collected in a bowl or container placed beneath the vehicle. Never use radiator coolant under any circumstances, for the reasons set out in Chapter 11.

Apply the natural sourcing techniques from Part III to the surrounding area without leaving sight of the vehicle. A dry creek bed within view, as covered in Chapter 7's seep well technique, or vegetation indicators from Chapter 8, can supplement the vehicle's own supply without requiring the traveller to leave the vicinity of the vehicle itself.

Signal deliberately. Raise the bonnet as a universal distress signal, use a mirror to flash sunlight toward any aircraft or distant vehicle, and activate a Personal Locator Beacon (PLB) or satellite phone if one is carried and the situation is assessed as genuinely serious, since this is by far the most effective way to bring rescue to a fixed location.

The preventive baseline matters more than any in-situation technique. Every credible source on outback travel converges on the same planning figure: carry at least 20 litres of water for the vehicle itself, plus an additional five to seven litres per person per day for the duration of the trip, and always advise someone of the planned route and expected return time before departing. A stranding with this baseline already in the vehicle is a manageable, survivable situation. A stranding without it is a genuine emergency from the first hour.

Shipwrecked on an Uninhabited Island

Isolation on an uninhabited coastline or island removes access to mains water entirely, but frequently provides more natural water sources than a desert stranding, provided the available techniques are applied systematically.

Assess the immediate coastline for a beach or dune well, using the technique set out in Chapter 7. A lens of fresh or brackish water frequently sits above the saltwater intrusion beneath a sandy shoreline, several metres back from the high tide line, and this is often the fastest source of usable water available on an otherwise dry island.

Check for coconut palms, particularly relevant to the tropical Pacific and the far northern Australian coastline covered in Chapter 10. A green, unripe coconut provides several hundred millilitres of naturally sterile liquid requiring no treatment.

Set up rain catchment immediately, using any available tarp, sail canvas, large leaves, or a dug depression lined with plastic sheeting or a poncho, since tropical and coastal regions frequently receive rainfall events that a prepared collection system can capture in useful volume even without a purpose-built tank.

Build a solar still, using the below-ground technique from Chapter 9, particularly where soil moisture or beach vegetation is available to increase yield, understanding the realistic, modest output covered in that chapter.

Treat every source found this way using the screening and treatment methods in Parts V and VI before drinking, with the specific exception of intact, undamaged coconuts, which do not require treatment as covered in Chapter 10.

Never drink seawater. Seawater contains approximately 35 grams of dissolved salt per litre, considerably more than the human kidney can safely excrete, meaning drinking seawater accelerates dehydration rather than relieving it, regardless of how severe the thirst.

A Disabled Vessel Adrift at Sea

A boat that has lost power or is otherwise unable to make way, with limited onboard drinking water and no functioning desalination equipment, faces a genuinely severe water constraint, since the surrounding environment is entirely seawater and offers none of the terrain-based or plant-based sourcing options available on land.

Ration the existing supply immediately and conservatively, applying the priority framework from Chapter 22, while working on alternative sourcing rather than waiting until the existing supply is exhausted.

Deploy a floating solar still if one is carried, using the purpose-built inflatable design referenced in Chapter 9, which produces meaningfully more water than an improvised alternative in a marine setting, in the order of 1.4 to 2.4 litres per day under reasonable conditions.

Collect rainfall on any available surface, including sails, tarps or awnings rigged to funnel water into a container, since even a brief tropical rain event can produce a genuinely useful volume if the vessel is set up to capture it before the rain arrives rather than improvising during the event itself.

A manual reverse osmosis desalinator, where carried as part of a life raft or vessel safety kit, is the single most effective tool available in this specific scenario, directly converting seawater into drinking water without dependence on sunlight, weather or fuel, and should be the first resource checked and deployed if present on board.

Never drink seawater, for the same physiological reason set out above, and never substitute urine for water intake, since this concentrates waste products the body is specifically trying to eliminate and provides no net hydration benefit.

Minimise physical exertion and sun exposure to reduce fluid loss through perspiration, using any available shade, and reserve movement and activity for essential tasks only.

Lost on a Multi-Day Hike or Bushwalk

Becoming lost or immobilised, such as by a fall or injury, during a hike in Australian mountains, bushland or rainforest, without a known route back to a reliable water source, combines the terrain-reading techniques from Part III with the specific discipline of staying findable.

Stop moving once genuinely lost, and stay in a single, visible location wherever practical, for the same reason set out for vehicle breakdowns above: a stationary person is dramatically easier to locate than one who continues moving unpredictably through the bush, and a PLB or mobile signal, where available, pinpoints a fixed location far more reliably than a moving one.

Apply the landscape-reading techniques from Chapter 8 without travelling far from the chosen location: converging animal trails, vegetation changes, and the sound or sight of birds at dawn and dusk can all indicate a nearby water source without requiring an extended, disorienting search.

Collect dew in the early morning, as covered in Chapter 9, which is often the most reliable atmospheric source available in vegetated Australian bushland and requires no equipment beyond a piece of clothing.

Follow drainage lines downhill where a source must be actively sought, since water and, frequently, walking tracks and eventually roads or waterways used by other people, tend to follow the same downhill paths, making a drainage line a reasonable general direction to move in if remaining stationary is genuinely not viable.

Treat any water found using the full screening and treatment approach in Parts V and VI, since an unfamiliar bushland water source carries the same unknown contamination risk as any other natural source covered throughout this book.

Plane Crash in Remote Terrain

A plane crash in remote desert, tundra, jungle or isolated mountain terrain, where survivors must wait days for rescue, combines elements of both the vehicle breakdown and the lost-hiker scenarios above, with the additional advantage that a crash site is generally larger, more visible and easier for search crews to spot than a person or a smaller vehicle.

Remain at or near the crash site, for the same reasoning that applies to a vehicle breakdown: the wreckage is the most visible and most likely target of any search effort, and moving away from it significantly reduces the likelihood of being found.

Use wreckage materials for water collection, including any available metal panelling, canopy material or fabric, which can be improvised into rain catchment surfaces or solar still components using the techniques in Chapter 9.

Apply the terrain-specific sourcing method relevant to the actual environment: the desert vehicle breakdown guidance above for an arid crash site, the coastal and island guidance above for a crash near a coastline, or the general landscape and dew-collection techniques from Part III for a temperate or mountain environment.

Ration any onboard emergency water supply immediately and conservatively, using the framework in Chapter 22, while establishing supplementary sourcing rather than waiting for the initial supply to run out.

Fleeing Through Wilderness While Evading Pursuit

A distinct and more constrained version of the lost-hiker scenario arises when the goal is not to be found. This applies to fleeing armed conflict, civil unrest, or any other situation requiring genuine evasion, and the standard "stay put and signal" guidance that applies to every other scenario in this chapter is directly reversed here.

Prioritise natural water sources over any approach to roads, farms, or visible human settlement, using the landscape-reading and plant-extraction techniques from Chapters 8 through 10, since approaching a visible structure or road to seek water directly exposes the traveller to exactly the contact being avoided.

Avoid signal fires, bright light sources at night, and any other visibility-increasing action that every other scenario in this chapter recommends, since the objective here is the opposite of being located.

Move and collect water during low-visibility periods, generally dawn, dusk or full darkness with adequate navigation ability, rather than in full daylight, and rest in concealed locations during the periods of highest visibility risk.

Favour water sources away from human activity indicators, including livestock trails, agricultural land and any sign of recent human presence, both because these locations carry the contamination risks covered in Chapter 8 and because they carry the additional risk of encountering other people.

Apply the full screening and treatment approach from Parts V and VI to any water obtained this way, since the pressure to move quickly and avoid detection does not reduce the biological risk carried by an untreated natural source, and an illness contracted while evading pursuit compounds an already dangerous situation considerably.

The Common Thread

Every scenario in this chapter draws on the same underlying toolkit: the landscape and plant-based sourcing methods from Part III, the screening discipline from Chapter 14, and the treatment methods from Part VI. What changes between scenarios is not the available technique but the surrounding constraint, whether that constraint is staying visible for rescue, staying hidden from pursuit, or working with only the materials a vehicle, a vessel or a crash site happens to provide. Recognising which constraint applies to the actual situation, and adapting the standard toolkit to it rather than abandoning it, is the practical skill this chapter is built to convey.

Sources referenced in this chapter

Royal Flying Doctor Service, Outback Survival Tips, flyingdoctor.org.au

Crikey Camper Hire, Your Guide to Surviving Western Australia's Outback

Sea to Summit, How to Not Die in The Outback

CHAPTER 25

When Infrastructure Is Attacked or Fails

Every scenario in this book so far has assumed that, somewhere beyond the household, a functioning system exists: a treatment plant, a distribution network, a supply chain restocking the shelves at the local hardware store. This chapter addresses the scenarios in which that assumption fails, not because a storm has passed through or a pipe has burst, but because the system itself has been deliberately targeted, digitally compromised, or has ground to a halt under the weight of conflict or collapse. These are the scenarios in which nobody can tell a household when the water will come back on, because the people who would normally know that answer no longer have one themselves. A household's stored reserve is not just a convenience in this category of event. It is the entire margin between managing the situation and being at its mercy, for however long that turns out to be, and that uncertainty about duration is precisely what should drive every decision in this chapter toward the most conservative, most self-sufficient response available.

Deliberate Sabotage and Terrorism Targeting Water Infrastructure

Water treatment plants, pumping stations and major distribution pipelines are recognised, explicit targets in modern conflict and terrorism planning, precisely because they are lifeline infrastructure that a population cannot function without and cannot quickly replace.

The realistic effect of sabotage is loss of pressure, loss of treatment, or contamination, not necessarily all three at once. A damaged pumping station may leave pipes intact but without pressure. A damaged treatment plant may leave pressure intact but without disinfection. Both failure modes have been documented in real incidents, and a household should not assume that "the tap still runs" means "the water is still safe."

Assume contamination until proven otherwise. In the aftermath of any confirmed or suspected deliberate attack on water infrastructure, treat all mains water as unsafe and rely entirely on the household's own stored and treated supply, following the boil-water discipline set out in Chapter 14, until an official authority confirms restoration of safe supply.

This is precisely the scenario the two-week and long-term storage tiers in Chapters 4 and 5 exist for. A household with a genuine multi-week reserve already in place is not dependent on knowing how long the disruption will last, because it has bought itself time regardless of the answer.

Cyberattacks on Automated Water Systems

Modern water treatment and distribution relies heavily on programmable logic controllers and automated systems to manage chemical dosing and pumping, and this automation has become a documented, active target.

Real, confirmed incidents have already occurred, including attacks on water authority infrastructure in Pennsylvania, Texas and Kansas in recent years, in which attackers gained control of pressure and dosing systems before operators switched to manual override to prevent contamination reaching the public supply. These incidents demonstrate both the genuine vulnerability of automated water infrastructure and the fact that, so far, human intervention has generally caught the problem before it reached consumers.

The household implication is the same as for physical sabotage: assume the possibility of compromised chemical dosing or pressure loss, and do not assume mains water is safe purely because it is flowing. A cyberattack affecting water infrastructure is frequently reported publicly with some delay, meaning a household relying solely on official notification may be drinking compromised water for longer than a household defaulting to its own stored and treated reserve at the first sign of any irregularity.

A household's own water security is entirely independent of this threat category. Chemical dosing systems, pumping automation and treatment plant control systems are all upstream problems a stored reserve and a personal treatment capability, covered across Parts II and VI, sidestep completely.

Prolonged Electricity Grid Failure

An extended grid failure, whether from infrastructure attack, extreme weather, or cascading technical failure, affects water supply through a specific and often underappreciated mechanism: most modern water systems, from municipal treatment plants down to a household's own rainwater tank pump or bore pump, depend on electricity to move water at all.

Know exactly which parts of the household's water system depend on grid power, including any electric pressure pump on a tank or bore, as covered in Chapters 6 and 7, and have a manual override or backup power plan for each one identified before it is needed.

Gravity-fed systems, established in advance, remove this dependency entirely. A tank or drum elevated on a stand, as covered in Chapter 5, continues to provide water by gravity regardless of how long grid power remains down.

Duration is genuinely unknowable in this scenario, and this is the category of disruption in which the extreme-tier planning from Chapter 23, long-term bulk storage, a licensed bore with a solar or manual pump option, and full-spectrum treatment capability, pays for itself most directly. A household planning only for the mild and moderate tiers should treat an extended grid failure as the clearest possible argument for building toward the extreme tier over time.

War, Siege and Civil Unrest

Armed conflict, a siege, or serious civil unrest can simultaneously cut off water deliveries, damage the physical infrastructure that would otherwise be repaired quickly, and make travel to a well, pipeline or treatment plant genuinely dangerous even where the infrastructure itself remains intact.

Physical access risk is a distinct problem from infrastructure damage, and must be planned for separately. A functioning well half a kilometre away is of no practical use to a household that cannot safely reach it, and this changes the calculation entirely from the peacetime version of the same problem: proximity to a known source is not the same as access to it.

Minimise movement and exposure when collecting or accessing water under these conditions. Where any household water source, a tank, a well or a stored reserve, exists on the property itself, it should be prioritised heavily over any source requiring travel through contested or unsafe areas, even where the on-property source is smaller or of lower quality and requires more intensive treatment as a result.

Duration in this category is frequently the least predictable of all the scenarios in this chapter. A siege or conflict-driven water disruption can resolve in days or persist for months, and a household in this situation should default to the most conservative rationing discipline from Chapter 22 from the very first sign of disruption, rather than assuming a short-term outcome and adjusting downward only once a longer duration becomes apparent.

Evading Authorities on a Remote or Abandoned Property

A related but distinct scenario involves a household or individual deliberately avoiding contact with authorities, whether police or another controlling body, while sheltering on a remote or abandoned rural property, where the water infrastructure that exists may be unfamiliar, uncertain, or only partially functional.

Assess whatever water infrastructure exists on the property using the same framework as any unfamiliar source: check for an existing rainwater tank using the inspection points in Chapter 6, check for a bore or well using the assessment approach in Chapter 7, and never assume any existing system is either functional or safe without direct inspection.

Avoid drawing attention through visible water-related activity, in the same way the evasion guidance in Chapter 24 applies to wilderness movement: an actively pumped bore, a newly rigged rainwater catchment, or frequent visible trips to a dam or tank can all be as visible as a signal fire to anyone specifically watching the property, and water security in this scenario needs to be balanced against the same concealment priorities as the movement itself.

Treat every source on an unfamiliar property with full screening and treatment, using Parts V and VI, since an abandoned property's water infrastructure has, by definition, gone without the maintenance covered throughout Parts II and III of this book, and should be assumed contaminated or degraded until proven otherwise.

Global or Regional Supply Chain Collapse

A collapse affecting the supply chains that deliver bottled water, treatment chemicals, pump components, replacement filters and the fuel needed to run generators and vehicles represents a slower-moving but potentially far more prolonged version of the scenarios above.

This scenario attacks the resupply side of the household's plan, not the water itself. A household's existing stored reserve, rainwater tank and treatment equipment continue to function; what disappears is the ability to replace consumables, chemical treatment tablets, filter cartridges, and fuel, once the current stock runs out.

Consumable-free and low-consumable methods become disproportionately valuable in this scenario. Boiling, provided a fuel source remains available, and a ceramic or ultra-long-life hollow fibre filter, as covered in Chapter 17, both continue to function without the ongoing chemical resupply that a tablet-based system depends on.

Stockpiling consumables ahead of time, at the extreme end of the preparedness tiers in Chapter 23, is the direct household-level defence against this scenario, and a household that has assessed its region as carrying genuine supply chain fragility risk should weight its treatment kit toward methods requiring the fewest ongoing consumable inputs.

The Unifying Principle: Plan for Unknown Duration

Every scenario in this chapter shares a single defining feature that distinguishes it from the storms, floods and blackouts covered in Parts I through VII: nobody, including the authorities who would normally provide an estimate, can reliably say how long the disruption will last. A flood recedes on a broadly predictable timeline. A cyberattack, a siege, or a supply chain collapse does not come with an estimated resolution date, and a household's response should be calibrated to that uncertainty rather than to an assumed best case. This is the practical argument, more than any other in this book, for building toward the extreme preparedness tier described in Chapter 23: not because every household will face one of these scenarios, but because the scenarios that do occur offer no reliable signal for when the need for self-sufficiency will end, and the household holding the largest independent margin, in storage, in sourcing capability and in treatment capacity requiring the fewest ongoing inputs, is the household best positioned to outlast whatever timeline actually unfolds.

Sources referenced in this chapter

CYFIRMA, U.S. Water Structure's Vulnerability to Cyber Attacks

Industrial Cyber, reporting on the Municipal Water Authority of Aliquippa and Tipton Municipal Utilities cyberattacks

NPR, Flushable Wipes and Iran: Water Treatment Facility Adds Cyberattacks to Worry List

Nextgov/FCW, reporting on the Muleshoe, Texas water facility breach

CHAPTER 26

When Nature Destroys the System

The scenarios in Chapter 25 involved deliberate or digital attack. This chapter covers the same category of outcome, water infrastructure destroyed or rendered unsafe rather than merely interrupted, arriving instead through the raw physical force of an earthquake, a cyclone, a flood, a volcanic eruption, a bushfire, a drought, or an extreme heatwave. These events share a brutal commonality with the infrastructure attacks in Chapter 25: nobody can promise a household how long it will take to rebuild a treatment plant that no longer exists, re-lay a water main that has been ripped apart by ground movement, or refill a dam that has not seen meaningful rain in years. A household's own water security, in every one of the scenarios below, is what stands in the gap between the disaster and the eventual, uncertain restoration of the system that failed.

Earthquake

An earthquake severe enough to rupture underground water mains does more than interrupt supply; it frequently contaminates whatever remains in the system with sewage, fuel or debris drawn in through the same breaks that let pressure escape.

Assume contamination, not merely absence, following any earthquake affecting the local water system. A cracked main does not simply leak; the loss of pressure allows soil, sewage from adjacent ruptured sewer lines, and any other surrounding contaminant to be drawn into the system through the same break, meaning water that continues to flow from a tap after a significant earthquake should be treated with the same suspicion as floodwater, using the boil-water discipline in Chapter 14, rather than assumed safe because it is still running.

The household's stored reserve, established in Chapters 3

through 5, is the only water supply not exposed to this specific contamination mechanism, since it sits isolated in a sealed container rather than connected to a network that has just been physically breached.

Shut off the household's own water supply at the main if a significant earthquake is confirmed and damage is suspected, both to prevent contaminated water entering the home's internal plumbing and to preserve whatever clean water already sits within the household's own pipes and hot water system, recoverable using the methods in Chapter 11.

Cyclone and Storm Surge

A cyclone combines high wind damage to treatment infrastructure with a second, distinct hazard: storm surge, which can drive seawater directly into freshwater sources, rendering them saline and unusable for an extended period even after the storm itself has passed.

Storm surge contamination does not resolve quickly. Saline intrusion into a coastal aquifer, a low-lying dam, or a damaged treatment plant's intake can persist for weeks to months after the surge itself has receded, since the underlying water source itself has been altered, not merely the infrastructure drawing from it.

A household's own elevated, sealed storage, established before the cyclone using the bathtub pre-fill and bulk storage methods in Chapters 3 through 5, is entirely unaffected by storm surge, provided it was filled and secured before the surge arrived rather than left exposed at ground level in a location the surge could reach.

Rainwater harvesting, covered in Chapter 6, becomes considerably more valuable rather than less in the aftermath of a cyclone, since the same weather system delivering the surge typically also delivers substantial rainfall, and a tank with intact first-flush and inlet screening continues to provide a genuinely fresh source even while the surrounding groundwater and surface water remain compromised by salt.

Flooding

Flooding overwhelms sewage systems in a way that makes local tap water unsafe far more often, and for far longer, than most households anticipate, and the pattern is well documented across Queensland and New South Wales flood events referenced throughout this book.

Assume mains water is unsafe for the full duration of any official boil-water notice, and beyond it if there is any reason to suspect the household's specific service line was affected differently from the wider network, as covered in Chapter 14.

Floodwater itself, wherever it has entered the property, should be treated as carrying sewage, agricultural chemicals, fuel and debris, regardless of how it appears. This applies to water in a flooded rainwater tank, which should be assumed contaminated and should be drained, cleaned and refilled once floodwaters recede, rather than simply resumed as a drinking source.

The household's sealed, elevated stored reserve is again the primary defence, precisely because it is the one part of the household's water system not directly exposed to the floodwater itself.

Volcanic Eruption

Volcanic ash covering reservoirs and roof-catchment systems introduces a specific contamination profile: fine particulate matter and, depending on the specific eruption, dissolved chemical compounds including fluoride and sulphur compounds at levels that can render both surface water and rainwater catchment systems unsafe.

Ash-affected rainwater tanks should not be used for drinking without testing, since the chemical profile of volcanic ash varies by eruption and cannot be assumed safe simply because the ash itself is not visibly present in the water.

Disconnect roof catchment systems from the tank during an active ashfall event where possible, and inspect and clean the catchment area thoroughly before resuming collection once the ashfall has ended.

This is a genuinely low-probability scenario for most of Australia, but remains relevant to travellers and any Australian household in the more limited regions where volcanic activity is a realistic, if rare, consideration, and the same general principle, disconnect, assess, and treat with caution rather than assume safety, applies regardless of location.

Bushfire

Bushfire damage to water infrastructure is more insidious than most households expect, because the danger does not end when the fire passes and the visible destruction stops. Real Australian and international experience shows that intense heat alone, without the pipe itself necessarily burning, can cause plastic water pipes to degrade and leach genuinely dangerous chemical contamination directly into the water system.

Documented incidents in California, following the 2017

Tubbs Fire and 2018 Camp Fire, found benzene, a known carcinogen, in drinking water at levels far exceeding safety standards, in some cases thousands of times above the legal limit, traced to heat-damaged plastic pipes and service lines rather than to the original water source itself. Laboratory research subsequently confirmed that plastic pipes can begin leaching benzene and other volatile organic compounds into water after heat exposure alone, at temperatures well below what is required to ignite the plastic outright.

This means a household's own plumbing, not just the municipal water source, can become the contamination point after a bushfire has passed nearby, even where the home itself was not destroyed. Any household returning to a property after a significant bushfire in the area should not assume mains or tank water is safe simply because the taps still run and the water looks and smells normal, since this specific contamination is chemical rather than biological and frequently has no obvious sensory warning sign in its early stages.

Water authorities in affected areas typically issue specific "do not drink" and "do not boil" notices following this kind of contamination, distinct from the standard boil-water notice covered in Chapter 14, because boiling concentrates rather than removes these particular volatile chemical compounds, in the same way covered for other hydrocarbon contamination in Chapter 16. These specific notices should be treated with the same seriousness as any other official water safety advisory and followed until formally lifted.

A sealed, stored reserve, filled before the bushfire threat arrived and physically isolated from the household's own plumbing, remains safe through this specific contamination pathway, since it was never connected to the affected pipework in the first place.

Rainwater tanks affected by ash and debris from a nearby bushfire, as opposed to the plumbing contamination above, should be inspected and, if necessary, drained and cleaned following the maintenance triggers set out in Chapter 6.

Drought

An extended drought differs from every other scenario in this chapter in one important respect: its onset is slow and its duration is often not merely unknown but genuinely indeterminate, sometimes persisting for years, and Australia's own climate record includes multi-year droughts significantly exceeding the planning assumptions built into much household preparedness.

A household's response to drought should begin well before any dam, reservoir or tank is close to empty, using the drought-specific sizing guidance in Chapter 6 for rural properties, planning storage against the region's longest recorded dry period rather than an average rainfall year.

Water restrictions imposed by authorities during drought are a genuine early warning signal, not merely an inconvenience, and a household's own conservation discipline from Chapter 22 should tighten in parallel with, or ahead of, official restrictions rather than waiting for mandatory enforcement.

Every sourcing method in Part III becomes progressively more important as a drought extends, since the household's own stored and harvested supply is, by definition, being drawn down faster than it is being replenished for the duration of the event.

Extreme, City-Wide Heatwaves

An unprecedented heatwave affecting an entire city compounds two problems simultaneously: it sharply increases water demand across the whole population at once, for drinking, cooling and hygiene, while placing genuine strain on treatment and distribution infrastructure that was not necessarily designed for sustained demand at this level.

Apply the full heat adjustment from Chapter 2 without exception during any officially declared extreme heat event, since this is precisely the condition those adjustments were built for, and the gap between the moderate-conditions planning figure and the actual heatwave requirement is exactly where household water plans fail if the adjustment is skipped.

Watch for early signs of system-wide strain, including reduced pressure, discolouration, or official communication about supply constraints, and treat these as a signal to activate the household's own stored reserve and conservation discipline proactively, rather than waiting for a formal restriction notice.

Vulnerable household members, covered in Chapter 20, face compounding risk during a city-wide heatwave specifically, since the same event that increases their individual water and cooling needs is also the event most likely to place strain on the wider system they might otherwise rely on.

The Pattern Across Every Natural Disaster in This Chapter

Every scenario above ends at the same conclusion, because the underlying mechanism is the same regardless of which natural force is doing the damage: a household's own sealed, stored water, established and secured before the event arrives, is the one part of its water security that no earthquake, cyclone, flood, eruption, bushfire, drought or heatwave can reach directly. Everything else, the mains connection, the local dam, the treatment plant, the rainwater tank left uncovered, remains exposed to whatever physical force the disaster brings, for however long it takes authorities to repair, retest and formally clear it for use again.

Sources referenced in this chapter

Chemical & Engineering News, How Wildfires Can Contaminate Drinking Water; and California Wildfires Caused Unexpected Benzene Contamination of Drinking Water

Natural Resources Defense Council, Wildfires Can Spark Widespread Contamination of Public Water Supplies*

The Conversation, Plastic Pipes Are Polluting Drinking Water Systems After Wildfire

PMC/National Institutes of Health, Organic Chemical Contaminants in Water System Infrastructure Following Wildfire

CHAPTER 27

Trapped and Displaced

This final chapter closes Part VIII, and this book, with two scenarios that share a single underlying feature despite looking nothing alike on the surface: the person or household affected has no meaningful control over their own water situation, because either the physical environment has sealed them in, or the sheer number of people sharing a limited water source has overwhelmed it entirely. In every earlier scenario in this book, from a household's own kitchen to a stranded vehicle in the outback, the person facing the crisis retained some agency over their own sourcing, treatment and rationing decisions. The two scenarios in this chapter strip much of that agency away, and the response has to be built around that constraint directly.

Trapped in a Collapsed Structure

Entrapment in a collapsed building, an underground structure, a mine, a tunnel or a bunker, with only whatever water happens to already be present, is among the most severe scenarios this book addresses, precisely because the trapped person or group typically cannot source, harvest or treat anything beyond what already exists within the confined space.

Inventory every drop of liquid immediately, without exception. This includes any beverage of any kind, not only water, since in a genuine entrapment scenario the ordinary caution against sugary or caffeinated drinks becomes secondary to the more urgent need for any fluid volume at all. Any liquid, however imperfect, extends survivable time.

Ration immediately and severely, applying the tightest end of the framework in Chapter 22 from the very first hour, since a trapped person has no way of knowing, and no reliable way of estimating, how long rescue will take. Real mine collapse and building collapse rescues have taken anywhere from hours to, in well-documented extreme cases, weeks, and there is no way to know in advance which timeline applies to a given incident.

Minimise physical exertion and, where the environment allows it, minimise body temperature elevation, since both directly increase fluid loss through perspiration and respiration, and in a fixed water supply situation, reducing consumption need matters as much as managing the supply itself.

Search the immediate environment systematically for any additional source, including condensation on cool surfaces such as concrete or rock, which can be collected in the same way as the dew and condensation techniques in Chapter 9, and any water contained within damaged plumbing, machinery cooling systems, or equipment specific to the structure, such as a mine's own water suppression or cooling systems, which may hold usable, if untreated, water.

Signal and make noise consistently but strategically, conserving energy between signalling attempts, since rescue teams in a structural collapse frequently rely on detecting sound or movement to pinpoint a trapped person's location, and the same energy conservation principle that applies to physical exertion generally applies to signalling effort as well: steady, periodic signals sustained over time serve better than a single, exhausting burst of effort followed by silence.

Treat any water obtained from an unfamiliar source within the structure with the same caution as any other unknown water source once free to do so, using the screening and treatment approaches in Parts V and VI, recognising that in the acute phase of entrapment itself, the immediate priority is survival to rescue, and any available fluid should generally be used rather than withheld over a treatment concern that is secondary to the more urgent need to survive until rescue arrives.

Mass Displacement and Refugee-Scale Demand

A refugee evacuation or mass displacement event, in which thousands of people converge on the same limited wells, tanks and emergency supplies, represents the inverse problem to entrapment: water may genuinely exist nearby, but the sheer number of people drawing on it simultaneously overwhelms the source's capacity, and an individual household's usual planning assumptions from Parts I through VII no longer hold in a shared, contested resource environment.

Understand the real, documented benchmark for this scenario. International humanitarian standards, set out by the Sphere Project and the United Nations High Commissioner for Refugees (UNHCR), specify a minimum of 15 litres per person, per day in the initial emergency phase and a target of 20 litres per person, per day for a protracted displacement situation. In practice, UNHCR's own reporting shows that this target is met in fewer than half of the settlements it supports globally, with some camps operating as low as 10 to 14 litres per person per day during periods of new arrivals or funding shortfall. This is the realistic benchmark to plan against in a genuine mass displacement scenario: considerably below normal household consumption, and not guaranteed even at that reduced level.

A household's own portable water security, established before displacement occurs, is disproportionately valuable in this scenario, since it is the only water supply not subject to sharing with an unknown and potentially very large number of other displaced people. A personal filter straw or squeeze filter, a small chemical disinfection supply, and a portable, carried container, as covered in Chapters 3 and 17, allow a displaced household to draw independently on whatever local source exists, a river, a well, a tank, rather than relying entirely on a shared distribution point.

Livestock and animals accompanying a displaced household add a further, frequently underestimated

demand. Documented humanitarian guidance allocates approximately 30 litres per day for a cow or camel and 15 litres per day for a goat or other small animal travelling with a displaced population, volumes that compete directly with human needs at an already strained shared source and should be factored explicitly into any household's displacement planning rather than assumed to be absorbed informally.

Assume any shared or communal water source in a mass displacement setting is contaminated until treated, regardless of its apparent source or the assurances of others at the site. UNHCR's own field guidance for exactly this scenario states plainly that water should be assumed dangerously contaminated unless proven otherwise by testing, precisely because large numbers of people sharing a single source under stress and with limited sanitation infrastructure is one of the most reliable conditions for a waterborne disease outbreak documented anywhere in public health literature.

Prioritise water quantity first, then quality, in the acute emergency phase, but never abandon quality entirely. Humanitarian guidance for this scenario explicitly notes that securing sufficient quantity takes practical priority in the first hours and days, since a person cannot survive on a small volume of perfectly clean water any better than a larger volume of adequately treated water, but this is a sequencing decision under acute pressure, not a licence to disregard treatment once the household has any capacity at all to apply it.

Where a family group is moving together, distribute both water and treatment capability across multiple members rather than concentrating it in one person or one container, so that the loss, theft or damage of any single item does not eliminate the household's entire water security at once, a specific and serious risk in a chaotic, crowded displacement setting.

The Overlap Between These Two Scenarios

Entrapment and mass displacement look like opposite problems, one person with too little water and no source, one shared source with too many people, but both point to the same practical lesson: whatever water security a person or household has already built for themselves before the crisis begins is worth disproportionately more than anything they can hope to arrange once the crisis has already begun. A trapped person cannot go shopping for a filter. A displaced family arriving at an overwhelmed camp cannot install a rainwater tank. The entire structure of this book, storage first, then sourcing, then treatment, then a written plan, exists because the moment to build water security is always before it is needed, and these final two scenarios represent the sharpest possible illustration of what happens when that moment has already passed.

Closing the Book

This book opened with a simple, deliberately blunt claim: the human body can survive roughly three weeks without food, but only three days without water, and that gap is the reason water deserves to sit at the very top of any preparedness plan rather than somewhere below torches, batteries and tinned food. Every chapter since has been an attempt to make that claim genuinely actionable, from calculating a household's real daily requirement, through storing it, harvesting it, finding it in places most people never think to look, treating it to a standard that actually protects health rather than merely feeling like it should, and finally, in this last part, applying every one of those skills to the extreme and improbable circumstances in which the ordinary rules no longer quite apply.

None of this requires turning a home into a bunker or a family into a unit of full-time preppers. It requires the same modest, repeatable discipline this book has returned to in every chapter: know the number, in litres per day, that the household actually needs; know where that water is coming from today, this week and this year; know how to make an uncertain source safe rather than merely hoping it is; and revisit all of it, deliberately, on the same recurring schedule set out in Chapter 23, so the plan stays current rather than becoming a document written once and forgotten. A household that has done this has done the single most important thing any preparedness effort can achieve: it has made the most urgent resource in any disruption the one thing it no longer has to worry about.

Sources referenced in this chapter

Sphere Association, The Sphere Handbook: Humanitarian Charter and Minimum Standards in Humanitarian Response

UNHCR, Water Manual for Refugee Situations; and Climate Change Threatens Water Supply for Refugees

SSWM (Sustainable Sanitation and Water Management), Water Purification in Emergencies Overview

PMC/National Institutes of Health, A Systematic Review of the Amount of Water per Person per Day Needed to Prevent Morbidity and Mortality in (Post-)Disaster Settings